Commercial Vehicle Safety
Technology and Practice in Europe

U.S. Department of Transportation
Federal Highway Administration

International Technology Exchange Program
May 2000

NOTICE

The contents of this report reflect the views of the authors, who are responsible for the facts and accuracy of the data presented herein. The contents do not necessarily reflect the official policy of the Department of Transportation.

The metric units reported are those used in common practice by the persons interviewed. They have not been converted to pure SI units because in some cases, the level of precision implied would have been changed.

The United States Government does not endorse products or manufacturers. Trademarks or manufacturers' names appear herein only because they are considered essential to the document.

The publication of this document was sponsored by the U.S. Federal Highway Administration under contract number DTFH61-99-C00005. awarded to American Trade Initiatives, Inc. Any opinions, options, findings, conclusions, or recommendations expressed herein are those of the authors and do not necessarily reflect those of the U.S. Government, the authors' parent institutions, or American Trade Initiatives, Inc.

This report does not constitute a standard, specification, or regulation.

Technical Report Documentation Page

1. Report No. FHWA-PL-2000-010	2. Government Accession No.	3. Recipient's Catalog No.
4. Title and Subtitle Commercial Vehicle Safety – Technology and Practice In Europe		4. Report Date May 2000
		6. Performing Organization Code:
7. Author(s) Kate Hartman, Bob Pritchard, Ken Jennings, Jim Johnson, Ron Knipling, John MacGowan, Larry Oliphant, Mike Onder, Charles Sanft		8. Performing Organization Report No.
9. Performing Organization Name and Address American Trade Initiatives P.O. Box 8228 Alexandria, VA 22306-8228		10. Work Unit No.(TRAIS)
		11. Contract or Grant No. DTFH61-99-C-0005
12. Sponsoring Agency Name and Address Office of International Program Office of Policy Federal Highway Administration U.S. Department of Transportation		13 Type of Report and Period Covered
		14. Sponsoring Agency Code
15. Supplementary Notes FHWA COTR: Donald W. Symmes, Office of International Programs		

16. Abstract

The United States and countries of the European Union share many of the same concerns and face similar challenges about commercial vehicle safety issues. This summary report describes the September 1998 technology transfer scan tour to four European countries to learn how these countries are addressing their own safety issues, even as they comply with the increasing centralization of rules and regulations enacted by the European Commission. The nine-member scan tour team that visited France, Germany, Sweden, and The Netherlands represented the FHWA, the Virginia and Minnesota State Departments of Transportation, the Owner/Operators and Independent Drivers Association, and independent transportation consultants. The key areas examined by the team were human resources, vehicle safety systems, and regulations. The report also includes recommendations and implementation strategies.

As this report demonstrates, team members were particularly interested in the European's integrated approach to driver training and preparation, the role of truck manufacturers in assessing crash causes and statistics to improve safety design, and the public/private partnerships that enhance training and safety, and augment regulatory policies and practices. The team believes that these practices provide valuable models for fresh opportunities for public/private cooperation in the areas of safety enhancement, regulatory policies, and standards enforcement for the U.S. motor carrier industry.

17. Key Words	18. Distribution Statement		
Key Words: hours-of-service regulations, driver education and training, driver simulators, CVO testing and licensing, data privacy, crash investigation, cab-crashworthiness, safety systems design, size and weight standards/rules, longer combination vehicles, in-company inspection, roadside inspection, automated data collection systems	No restrictions. This document is available to the public from the Office of International Programs FHWA-HPIP, Room 3325 US Dept. of Transportation Washington, DC 20590 international@fhwa.dot.gov www.international.fhwa.dot.gov		
19. Security Classif. (of this report) Unclassified	20. Security Classif. (of this page) Unclassified	21. No. of Pages 60	22. Price

Form DOT F 1700.7 (8-72)   Reproduction of completed page authorized

COMMERCIAL VEHICLE SAFETY TECHNOLOGY AND PRACTICE IN EUROPE

Kate Hartman
FHWA
(Chairperson)

Bob Pritchard
Cambridge Systematics
(Report Facilitator)

Ken Jennings
Virginia DOT

Jim Johnson
Operators and Independent Drivers Assoc.

Ron Knipling
FHWA

John MacGowan
FHWA

Larry Oliphant

Mike Onder
US DOT

Charles Sanft
Minnesota Department of Transportation

and

American Trade Initiatives, Inc.

&

Avalon Integrated Services, Inc.

for the

Federal Highway Administration
U.S. Department of Transportation
Washington, D.C. 20590

May 2000

ii

FHWA INTERNATIONAL TECHNOLOGY EXCHANGE PROGRAMS

The FHWA's international programs focus on meeting the growing demands of its partners at the Federal, State, and local levels for access to information on state-of-the-art technology and the best practices used worldwide. While the FHWA is considered a world leader in highway transportation, the domestic highway community is very interested in the advanced technologies being developed by other countries, as well as innovative organizational and financing techniques used by the FHWA's international counterparts.

INTERNATIONAL TECHNOLOGY SCANNING PROGRAM

The International Technology Scanning Program accesses and evaluates foreign technologies and innovations that could significantly benefit U.S. highway transportation systems. Access to foreign innovations is strengthened by U.S. participation in the technical committees of international highway organizations and through bilateral technical exchange agreements with selected nations. The program is undertaken cooperatives with the American Association of State Highway Transportation Officials and its Select Committee on International Activities, and the Transportation Research Board's National Highway Research Cooperative Program (Panel 20-36), the private sector, and academia.

Priority topic areas are jointly determined by the FHWA and its partners. Teams of specialists in the specific areas of expertise being investigated are formed and sent to countries where significant advances and innovations have been made in technology, management practices, organizational structure, program delivery, and financing. Teams usually include Federal and State highway officials, private sector and industry association representatives, as well as members of the academic community.

The FHWA has organized more than 35 of these reviews and disseminated results nationwide. Topics have encompassed pavements, bridge construction and maintenance, contracting, intermodal transport, organizational management, winter road maintenance, safety, intelligent transportation systems, planning, and policy. Findings are recommended for follow-up with further research and pilot or demonstration projects to verify adaptability to the United States. Information about the scan findings and results of pilot programs are then disseminated nationally to State and local highway transportation officials and the private sector for implementation.

This program has resulted in significant improvements and savings in road program technologies and practices throughout the United States, particularly in the areas of structures, pavements, safety, and winter road maintenance. Joint research and technology-sharing projects have also been launched with international counterparts, further conserving resources and advancing the state-of-the-art.

For a complete list of International Technology Scanning topics, and to order free copies of the reports, please see the last page of this publication.

Website: www.international.fhwa.dot.gov
E-Mail: international@fhwa.dot.gov

ACRONYMS

ACC	automated cruise controls
AFT-IFTIM	Association for the Development of Professional Training in Transport — Institute of Training and Warehousing Techniques (France)
ATA	American Trucking Association
BASt	Federal Highway Research Institute (Germany)
CBR	Centraal Bureau Rijvaardigheidsbewizen (The Netherlands)
CVS	Commercial Vehicle Safety
EC	European Commission
EU	European Union
FHWA	Federal Highway Administration
FHWA/OMCHS	Federal Highway Administration Office of Motor Carrier and Highway Safety. December 1999 legislation separated the two programs and established the Federal Motor Carrier Safety Administration (FMCSA). Safety is a Core Business Unit within the Federal Highway Administration.
GPS	Global Positioning System
INRETS	National Institute for Transport and Safety Research (France)
ISA	intelligent speed adaptation
LCVs	longer combination vehicles
NHTSA	National Highway Traffic Safety Administration
OEM	original equipment manufacturer
TNO	Organization for Applied Scientific Research (The Netherlands)
TYA	Vocational Training and Working Environment Council (Sweden)
U.S. DOT	United States Department of Transportation

CONTENTS

EXECUTIVE SUMMARY vii

CHAPTER ONE – INTRODUCTION 1
    ENHANCED SAFETY GOALS 1
    CVS PANEL METHODOLOGY 2
    CVS PANEL AND SPONSORING AGENCY 2

CHAPTER TWO – HUMAN RESOURCE MANAGEMENT 4
    FOCUS AREAS 4
        Education and Training 4
        Testing and Licensing 7
        Hours of Service Regulations 8
        Onboard Recorders and Data Privacy 8
        Motor Carrier Approach to Drivers 10
    FINDINGS AND RECOMMENDATIONS 11
    IMPLEMENTATION STRATEGY 12

CHAPTER THREE – VEHICLE SAFETY SYSTEMS 14
    FOCUS AREAS 14
        Safety Research Business Models 14
        Crash Investigation 17
        Vehicle and Safety Systems Design 19
        Onboard Safety 22
    FINDINGS AND RECOMMENDATIONS 23
    IMPLEMENTATION STRATEGY 24

CHAPTER FOUR – SAFETY REGULATIONS AND ENFORCEMENT 26
    FOCUS AREAS 26
        Rules and Regulations 26
        Enforcement Models 27
        Information Systems 31
    FINDINGS AND RECOMMENDATIONS 32
    IMPLEMENTATION STRATEGY 33

**CHAPTER FIVE – CONCLUSIONS - ADVANCING THE DRIVER,
VEHICLE, AND REGULATIONS FOR ENHANCED SAFETY** 35
    UNDERSTANDING THE EUROPEAN APPROACH 35
    Safety is the Absence of Failure 35
    The Driver 35
    Integration 36
    DEPLOYMENT STRATEGY 36

APPENDIX A – AMPLIFYING QUESTIONS 37

APPENDIX B – COMPOSITE AGENDA 39

APPENDIX C – CVS PANEL BIOGRAPHICAL INFORMATION 41

**APPENDIX D – CONTACTS/WEB ADDRESSES OF PARTICIPANTS AND
ORGANIZATIONS** 44

FIGURES

Figure 1. CVS Panel Conclusions: Advancing the Driver and Vehicle for Enhanced Safety ix
Figure 2. CVS Panel Research Methodology 3
Figure 3. Driver Education at Stora Holm, Gothenberg, Sweden 5
Figure 4. Driver Education Promotion from the Vocational Training and Working Environment Council, Sweden 7
Figure 5. Tachograph Disk 9
Figure 6. Crash Investigation Results from Volvo 18
Figure 7. Integrated System from DaimlerChrysler 21
Figure 8. The Dutch Enforcement Model 28

EXECUTIVE SUMMARY

A presentation by Dr. Reinhard Ball of DaimlerChrysler that forecast future freight transportation in Germany could easily have described the situation in several U.S. States – 45 percent increase in truck ton-miles by 2010, rapidly growing passenger traffic, and flat infrastructure investments. Europe and the United States both experience increasing numbers of trucks registered, vehicle miles traveled, and a high number of fatal crashes.

The United States and Europe share common commercial vehicle safety issues, including a debate over access for longer combination vehicles; a shortage of commercial vehicle drivers; the need to integrate emerging public and private information technology systems; emerging technologies and developments in areas of safety systems that necessitate new standards; the emergence of rules/regulations from a centralized government with a decentralized enforcement approach.

Europe and the United States are also approaching commercial vehicle safety issues in similar ways that strive to strengthen the relationships between driver; vehicle, rules, and regulations; and the supporting organizations and institutions.

To explore ways to improve commercial vehicle safety on America's roadways, the Federal Highway Administration's International Technology Exchange Program convened the Commercial Vehicle Safety (CVS) Panel. The Panel focused its research on four European countries – Sweden, Germany, the Netherlands, and France. Panel members were Kate Hartman (Chairperson), Federal Highway Administration (FHWA); Bob Pritchard (Report Facilitator), Cambridge Systematics; Ken Jennings, Virginia Department of Transportation; Jim Johnston, Owner/Operators and Independent Drivers Association; Ron Knipling, FHWA; John MacGowan, FHWA; Larry Oliphant, Transportation Consultant; Mike Onder, U.S. DOT; and Charles Sanft, Minnesota Department of Transportation.

Note: The Motor Carrier Safety Improvement Act, which was signed into law on December 9, 1999, established a new Federal Motor Carrier Safety Administration (FMCSA) within the U.S. Department of Transportation. Prior to that, the Federal Highway Administration administered the Office of Motor Carrier and Highway Safety Program. The mission of the FMCSA is to improve truck and bus safety on our Nation's highways through information technology, targeted enforcement, research and technology, outreach, and partnerships.

The CVS Panel considered many emerging safety systems, technologies, and issues; it also developed recommendations for enhancing commercial vehicle safety in the United States. In preparation for the scan tour, panel members submitted introductory queries to the host governments. In turn, each host nation arranged visits and scheduled tours and discussions with regulatory and roadway operations agencies, research organizations, equipment manufacturers, and freight transportation firms. Sessions focused on the areas of human factors, equipment, infrastructure, and organizational structures.

The CVS Panel and this report concentrate on investigating approaches to enhancing safety of the commercial vehicle driver, the performance of the vehicle, and the accompanying rules and regulations. Uniting these three areas are the organizations that make it work – the innovative relationships that are central to understanding and new approaches and assimilating findings.

This report presents the panel's findings and proposes recommendations to support the strategies that advance human resource management, vehicle and roadway safety design and standards, and innovative regulatory methods to facilitate adoption of new technologies and approaches.

HUMAN RESOURCE MANAGEMENT

Driver management in Europe begins with mandatory and extensive training in the form of proactive education. Motor carriers are identifying and recruiting drivers to make full use of driving time constrained by mandatory by hours-of-service regulations. New flows of information in the vehicle and to the driver will require new skills and challenge the abilities of drivers in the future. The Europeans are addressing these with initiatives to educate drivers and create business competency.

In the United States, areas for advancement include:

- Driver education, specifically by developing a well-rounded, standard curriculum.
- Performance-based driver assessment, using performance data to better understand the needs of drivers and carriers and to develop public policy.

VEHICLE SAFETY SYSTEMS

Truck manufacturers have developed and are deploying many new vehicle safety systems – from airbags to collision-avoidance systems. Their approach is built on the needs of the vehicle operators and designed with extensive testing and crash analysis. These safety systems are constrained by the needs to safely and effectively deliver information to drivers and to demonstrate a positive return on investment for company owners.

Areas for advancement in the United States include:

- Develop systems standards – notably, cab-crashworthiness, human-machine interface, and other relevant standards.
- Use crash investigation for vehicle design.
- Focus on user acceptance of safety systems to ensure maximum use.

REGULATIONS AND ENFORCEMENT

Within the European Union (EU), the European Commission (EC) is charged with creating safety regulations that apply throughout the member nations. Integrating national rules and regulations has created an interesting context for new regulatory development and for national regulatory enforcement efforts. There are several noteworthy models of safety compliance approaches, such as the recent Dutch innovation that combines roadside and in-company inspections.

This central EC role has evolved from assimilating existing national regulations (which continues as new nations join the EU) to creating new laws in many complex and technical areas. The EC attempts to establish and ensure compliance with its regulations without financial incentives for its member nations.

Many operators within the U.S. motor carrier industry have safety programs that meet or exceed the level of regulatory requirements. A clear understanding of

Driver 	Vehicle 	Regulations & Enforcement 
Human Resource Management	**Vehicle Safety Systems**	**Government Structures**
• Education	• Accident Investigation	• Rules and Regulations
• Testing and Licensing	• Vehicle and Safety System Design	• Enforcement Model
• Hours of Service Regulations	• Safety Research Business Models	• Information Systems
• On-Board Recorders	• On-Board Safety Systems	
• Optimization – Motor Carrier Approach to Driver Management		

FIGURE 1
CVS Panel Conclusions:
Advancing the Driver and Vehicle for Enhanced Safety

these leading practices by regulatory agencies may allow for self-certification or an annual in-company inspection process.

Areas for advancement in the United States include:

- Alternative and complementary inspection activities that focus on understanding how to augment motor carrier safety programs to allow for possible self-certification of motor carrier safety systems.

- Improved use of in-company inspections and third-party advisors to improve motor carrier regulatory compliance and allow government resources to be focused on high-risk carriers.

CONCLUSIONS – ADVANCING THE DRIVER AND VEHICLE FOR ENHANCED SAFETY

In many European countries, a comprehensive policy objective drove commercial vehicle safety programs. In the United States, the renewed momentum of the U.S. DOT can be enhanced with approaches to create an imperative to advance commercial vehicle safety. These efforts are:

1) Establish a "Safety Forum" comprised of public and private interests to discuss and seek resolution to today's safety issues; and

2) Convene a national safety conference to focus on the issues raised by the CVS Panel.

The United States has many noteworthy efforts to enhance commercial vehicle safety that have created many safety initiatives, including programs, organizational relationships, and emerging technology-based safety solutions. New initiatives to enhance safety in the U.S. should begin with these current efforts and encompass the examples from the European experience.

Each of the three areas of this report – human resources, vehicle safety systems, and regulations – identifies lead organizations and suggests implementation strategies.

Chapter One
Introduction

ENHANCED SAFETY GOALS

An important goal of the U.S. Department of Transportation (U.S. DOT) is to decrease the numbers and severity of crashes on roadways across our Nation. An important component of this goal is identifying ways to improve the safe transit of the millions of commercial vehicles that conduct the Nation's daily commerce. While commercial vehicle safety is improving, issues and challenges continue to increase as the pressures to move more people and goods mount. The bottom line, however, is that more than 40,000 individuals die each year on American roadways. About 5,000 of those fatalities involve commercial motor vehicles.

Efforts to improve safety on the Nation's roadways include the 1995 U.S. DOT-sponsored Truck and Bus Safety Summit, which called for reducing the number and severity of commercial vehicle crashes. Among the safety issues identified at the Summit were driver fatigue, the need for crash-cause data, technology, and uniformity in truck safety regulations.

Secretary of Transportation Rodney Slater articulated an ambitious goal in 1999 when he established the goal to decrease truck-related fatalities by 50 percent in the next 10 years. This new challenge is supported with increased penalties for safety violations and proposed new funds for additional safety inspections.

As the United States explores ways to improve commercial vehicle safety, our European counterparts are tackling the same problem. For example, safety regulators in the Netherlands cite the disproportionate number of fatalities involving heavy-duty vehicles (6.3 percent of vehicle kilometers – 3.8 percent serious injuries, 4.6 percent of all injuries, and 14.6 percent of fatalities). The Europeans, however, are addressing the problem with a clear understanding of the relationships between driver, vehicle, rules and regulations, and the supporting organizations and institutions. Mr. G.H. Doornick, from the Netherlands, summarizes the Dutch approach:

> "A sustainable safe road traffic system is one in which the road infrastructure has been adapted to the limitations of human capacity through proper road design, in which vehicles are technically equipped to simplify driving and to give all possible protection to vulnerable human beings, and where necessary, deterred from undesirable or dangerous behavior. Man should be the reference standard and road safety problems should be tackled at its roots."

This approach is borne out through innovative safety initiatives throughout the countries of the European Union (EU). Examples include the Dutch government *Sustainable Safety* initiative, the Swedish *Vision Zero*, and the current French commitment to reduce traffic fatalities 50 percent by 2003.

CHAPTER 1

European truck manufacturers are playing a significant role in improving motor carrier safety. Manufacturers have allocated resources to develop and test the engineering skills that can mitigate many safety problems. DaimlerChrysler pointed to the potential of drastically reducing crashes through active safety systems. Volvo has identified innovative new passive safety systems to protect the driver.

The CVS Panel research and findings build on Secretary Slater's challenge and the momentum generated at the Truck and Bus Safety Summit and the safety issues it identified – driver fatigue, the need for crash-cause data, driver training, technology, and uniformity in truck safety regulations. The panel focused on these safety issues and used the European experience to generate numerous recommendations for action.

CVS PANEL METHODOLOGY

The CVS Panel prepared for its September 1998 tour by posing a series of questions, assessing answers, and preparing recommendations based on commercial vehicle safety systems in Europe.

The questions addressed the three broad areas of human factors, equipment and infrastructure, and institutional context/background. The panel forwarded these questions to the governments of Sweden, Germany, the Netherlands, and France. In response, the lead government agencies arranged meetings that included public and private sector specialists in commercial vehicle safety. Panel members visited various offices and institutes and held in-depth discussions with representatives from each country's regulatory and roadway operations agencies, research organizations, equipment manufacturers, and freight transportation firms.

CVS PANEL AND SPONSORING AGENCY

The CVS Panel members were:

Kate Hartman (Chairperson), Transportation Specialist
Federal Highway Administration (FHWA)

Bob Pritchard (Report Facilitator), Transportation Consultant
Senior Associate, Cambridge Systematics

Ken Jennings, Maintenance Division
Virginia Department of Transportation

Jim Johnston, President
Owner/Operators and Independent Drivers Association

Ron Knipling, Chief, Research Division
FHWA Office of Motor Carrier and Highway Safety

John MacGowan
FHWA

Larry Oliphant
Truck Manufacturing Expert and Transportation Consultant

Mike Onder, US DOT
ITS Joint Program Office

Charles Sanft, Director, Freight Planning and Development
Minnesota Department of Transportation

Amplifying Questions → Sent to the federal transporta-
See Appendix A   tion agency in each country

Visits and Discussions → Generated a large volume of
See Appendix B   notes and bibliography

Three Meetings of the CVS Panel → Brought together the skill and
See Appendix C   expertise of the assembled
                                  group

Report Review and Production

The CVS Panel toured the four countries from September 12, 1998 to September 26, 1998. Three CVS Panel group meetings were held (September 12 in Gothenburg, September 19 in the Hague, and September 26 in Paris) and focused the findings, formulated recommendations, and developed a strategy for implementation.

FIGURE 2

CVS Panel Research Methodology

The Federal Highway Administration sponsored this public/private exploration through the International Technology Exchange Program. The American Trade Initiatives Company provided logistical support and guidance.

Please note that the Motor Carrier Safety Improvement Act, which was signed into law on December 9, 1999, established a new Federal Motor Carrier Safety Administration (FMCSA) within the U.S. Department of Transportation. Prior to that, the Federal Highway Administration administered the Office of Motor Carriers and the Motor Carrier and Highway Safety Program. The mission of the FMCSA is to improve truck and bus safety on our Nation's highways through information technology, targeted enforcement, research and technology, outreach, and partnerships.

Chapter Two

HUMAN RESOURCE MANAGEMENT

On both sides of the Atlantic, effective, successful motor carrier operations and freight flow rely on commercial vehicle drivers. In Europe, there is also a shortage of trained drivers and truck driving has a somewhat tarnished image as an occupational choice.

In general, Europeans take a holistic approach toward drivers that combines mandatory, cooperative, and comprehensive education and training with focused carrier selection and driver recruitment. The regulatory systems are prescriptive and are built upon the social and industrial structures of the respective countries. For example, there are hours-of-service restrictions, physical requirements, mandatory use of onboard recorders (tachographs), and driver-pay requirements (minimum salary and payment by the hour).

FOCUS AREAS

The CVS Panel's focus in Human Resource Management encompasses:

- Education and Training.
- Testing and Licensing.
- Hours-of-Service Regulations.
- Onboard Recorders and Data Privacy.
- Optimization – Motor Carrier Approach to Driver Management.

Education and Training

The basic driver training model in Europe is vocational education. The process is selective and advances overall business competence as well as driving skills. This contrasts with U.S. training – which is neither standardized nor mandatory – that focuses solely on developing driver skills.

In Europe, a public/private effort has established a standardized curriculum that often uses advanced technologies such as simulators and password-protected Internet access. Many government regulations are designed to protect drivers by shielding them from unsafe work conditions. Promotional activities publicize truck driver and related occupations.

Organizational Models

The CVS Panel visited two institutions that exemplify the European focus on using new technologies and related research to educate commercial vehicle drivers – the Association for the Development of Professional Training in Transport-Institute of Training and Warehousing Techniques (AFT-IFTIM) in Menchy Saint-Eloi, France; and Stora Holm in Gothenburg, Sweden.

The French Model

The ATF-IFTIM trains 20 percent of France's new commercial drivers each year. The institute is funded 75/25 percent by the private/public sectors. It is developing and deploying the new generation of driver training simulators and onboard recorders and has well-established physical and psychological requirements for student acceptance.

The institute offers an extensive curriculum that combines simulator, personal computer, and behind-the-wheel training. It also requires a prescribed number of classroom hours and closely monitors behind-the-wheel performance. Deploying driver simulators defrays in-cab training costs and allows for training in all weather. The institute uses an innovative onboard recording device (Pilote 2001) with individualized smart cards to establish driver trainee baseline performance and skills and to assess progress at prescribed intervals.

France funds these types of educational and training activities through a payroll tax (0.5 percent) collected from all employees. In addition to vocational training, drivers must participate in refresher programs every 5 years. France also has two noteworthy regulatory structures designed to advance the interests of the drivers. First, drivers are paid by the hour (payment based on distance traveled or value of load is considered unsafe and not allowed), and second, truck movements are not allowed on Sunday.

The Swedish Model

The Swedish approach concentrates on developing the overall competence of professional drivers. The approach is based on the belief that greater breadth in training will produce more effective, safer drivers. A list of attributes identified for screening potential drivers for occupational aptitude includes technical skill, punctuality, and safety consciousness.

FIGURE 3

Driver Education at Stora Holm, Gothenberg, Sweden

CHAPTER 2

The Vocational Training and Working Environment Council (TYA) is an organization of private sector employer and employee groups that includes trade associations and unions. Since 1975, this self-funded group has advanced driver-training requirements and promoted vocational training centers across Sweden. The Stora Holm in Gothenburg, Sweden, is an example of a municipal vocational center that offers the standard vocational curriculum and training program. Drivers take a 10-week course to qualify for a commercial vehicle license. Learning involves a combination of computers, simulators, and behind-the-wheel-training. A noteworthy advance is the use of computer-based training delivered via an extranet. The extranet training includes narrative Q&A and illustrations that use video stream inserts. All Stora Holm graduates are hired directly into the transport industry.

Both the Swedish and French education and training organizations have undertaken efforts to promote the commercial driver occupation and to improve its public image. While educational efforts are designed to improve the skills and business competence of commercial vehicle drivers, promotional activities describe their professionalism and detail the role that drivers play in society. In addition to expanding the pool of potential commercial drivers, promotional efforts are also designed to increase the pool of candidates.

Driver Simulators

Throughout Europe, driver simulators are becoming an important enhancement for cost-effective, safe driver training. They are cost-effective because they allow year-round training and cost less than behind-the-wheel training. Because simulators cannot capture real-life terrain and vehicle dynamics, the optimal blend of simulator/computer/behind-the-wheel training needs has not yet been determined. New systems are being developed and deployed, and developments in virtual reality and systems optimization promise more effective use of simulators.

> [The French Association for the Development of Professional Training...] considers 1 hour on the simulator and 4 hours behind the wheel to be more effective [training] than 8 hours behind the wheel.

Both Stora Holm and AFT-IFTIM use emerging systems. Stora Holm is developing its own system and Thomson Training and Simulation is leading the development of the French system. In France, the government pays the greater part, while in Sweden, the driver and/or company pays for the development and use of the simulators.

First-year deployment of the AFT-IFTIM's driver simulator yielded impressive results. Reports indicate both timesaving and training effectiveness. Most notable was enhanced maneuvering training. AFT-IFTIM considers 1 hour on the simulator and 4 hours behind the wheel to be more effective than 8 hours behind the wheel. The central issues for the future are who benefits and who pays. The benefits are spread across several groups, including the training institutions, the driver, the trucking company, and the motoring public.

CHAPTER 2

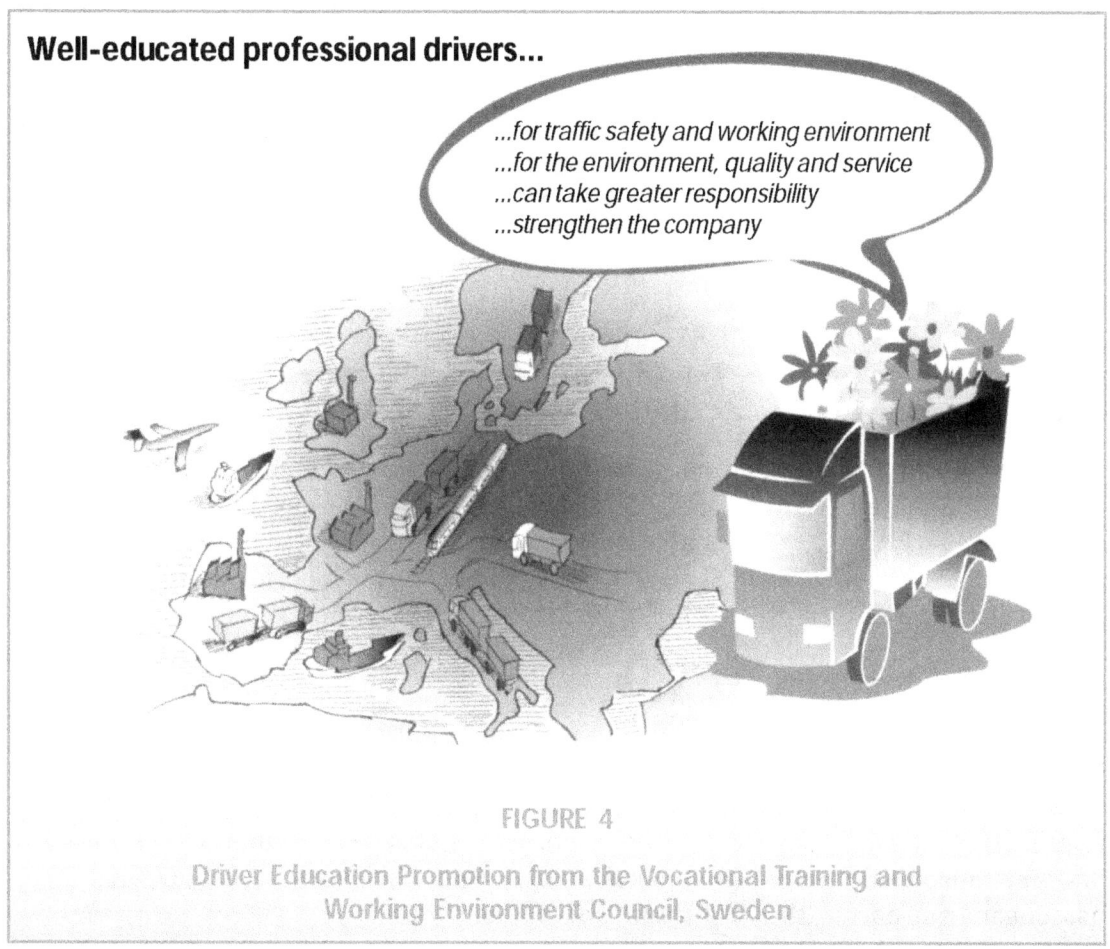

FIGURE 4

Driver Education Promotion from the Vocational Training and Working Environment Council, Sweden

Testing and Licensing

The European Commission (EC) codifies rules and regulations related to drivers. Each driver's home country issues a license that also allows for operation throughout the European community. The EC standardizes the testing process and each country deploys the testing procedure. Commercial licensing requires that drivers successfully pass both administrative (demonstrating knowledge of traffic rules, safety regulations, and vehicle mechanics) and vehicle operations tests. Government-certified organizations conduct most testing. In France, the AFT-IFTIM conducts the tests in addition to providing education.

The Dutch Model

The Dutch Centraal Bureau Rijvaardigheidsbewizen (CBR) is also a noteworthy organizational model. CBR is a driver testing center and an

organization that certifies professional driver competence. Since 1927, this private organization has maintained a meticulous admittance policy based on driving aptitude and skill, which has advanced the skill level of the truck driver population.

CBR sets and administers tests and provides value-added training and certification services to employers. In addition to the professional certificates required by the EC, CBR provides employers with extra, more rigorous certification, such as the ability to couple and uncouple trailers and semitrailers safely at a loading platform. CBR also provides advisors to motor carriers to help improve safety performance. This proactive approach of intervention also is practiced in other European countries.

Throughout Europe, selection for training and employment involves medical and psychological testing. As indicated above, there is a focus on driving aptitude by candidates. In France, there is also government-sponsored medical and/or psychological test of current drivers who demonstrate poor performance; the testing helps identify a program of treatment to improve performance.

Hours-of-Service Regulations

The most common commercial vehicle infraction in Europe, as in the United States, is violating hours-of-service regulations. In 1985, the EC amended the hours-of-service regulations originally set in 1969. That same year the EC also amended the related 1969 requirement for a tachograph. Hours-of-service regulations apply to drivers transporting freight in vehicles greater than 3.5 metric tons (4.6 tons) and passenger transport with more than nine passengers.

European regulations are as complicated as those in the United States. Europeans regulations allow for 45 driving hours per week averaged over 2 weeks – a maximum of 9 hours per day, except twice a week when 10 hours is permitted, if followed by 11 consecutive hours of rest. Every 4.5 hours of work must include 45 minutes of break time in segments of at least 15 minutes. Drivers must be compensated for all compulsory rest time, and there is an obligatory rest of 45 consecutive hours for every 6 days of work.

The EC recently debated the reducing the workweek to 35 hours. The outcome of this debate will particularly affect companies with more than 20 drivers. In the past, "other work time" was as high as 60 hours total. The recent debate included a proposal to reduce the total hours from 60 to 48. The new ruling from the EC is forthcoming.

Onboard Recorders and Data Privacy

The mechanical rotary tachograph is the primary tool for enforcing hours-of-service regulations. As noted above, it has been required since 1985 and is intended to protect the driver from abusive and unsafe working conditions. Tachograph tracks five activities: driving time; other work time such as truck washing and administrative work; availability time, for instance waiting for a truck to be unloaded/loaded; rest time; and 15-minute break time pauses. This mechanical device will be phased out in the coming years and replaced by an electronic device.

The current tachograph has been criticized for its susceptibility to large-scale fraud, enforcement difficulty, and costly company management. The emerging electronic device will not eliminate all criticism, but it should decrease the level of burden placed on motor carriers and drivers. A number of technology questions remain, for example, how to accurately track driver hours and not just vehicle operation. There is also a perceived need for a cost-effective onboard printer.

The vendors of these technologies suggest great promise for both regulatory improvements (good reporting capability including exceeding-the-speed limiter and other faults) and operational enhancements (more effective routing and dispatching and greater information flow). Several vendors offer electronic tachograph systems. Two vendors, Mannesmann VDO and Thomson, participated in CVS Panel discussions. Mannesmann VDO has also acquired interesting experience with its UDS (Accident Data Recorder), a black box system specially designed to register and monitor crash data. Crash rates, as well as expenditures for repair costs, could be reduced in several fleet trials, for example, with the Berlin police. Another Mannesmann VDO subsidiary, Mannesmann Passo GmbH, provides for a complete range of traffic telemetrics.

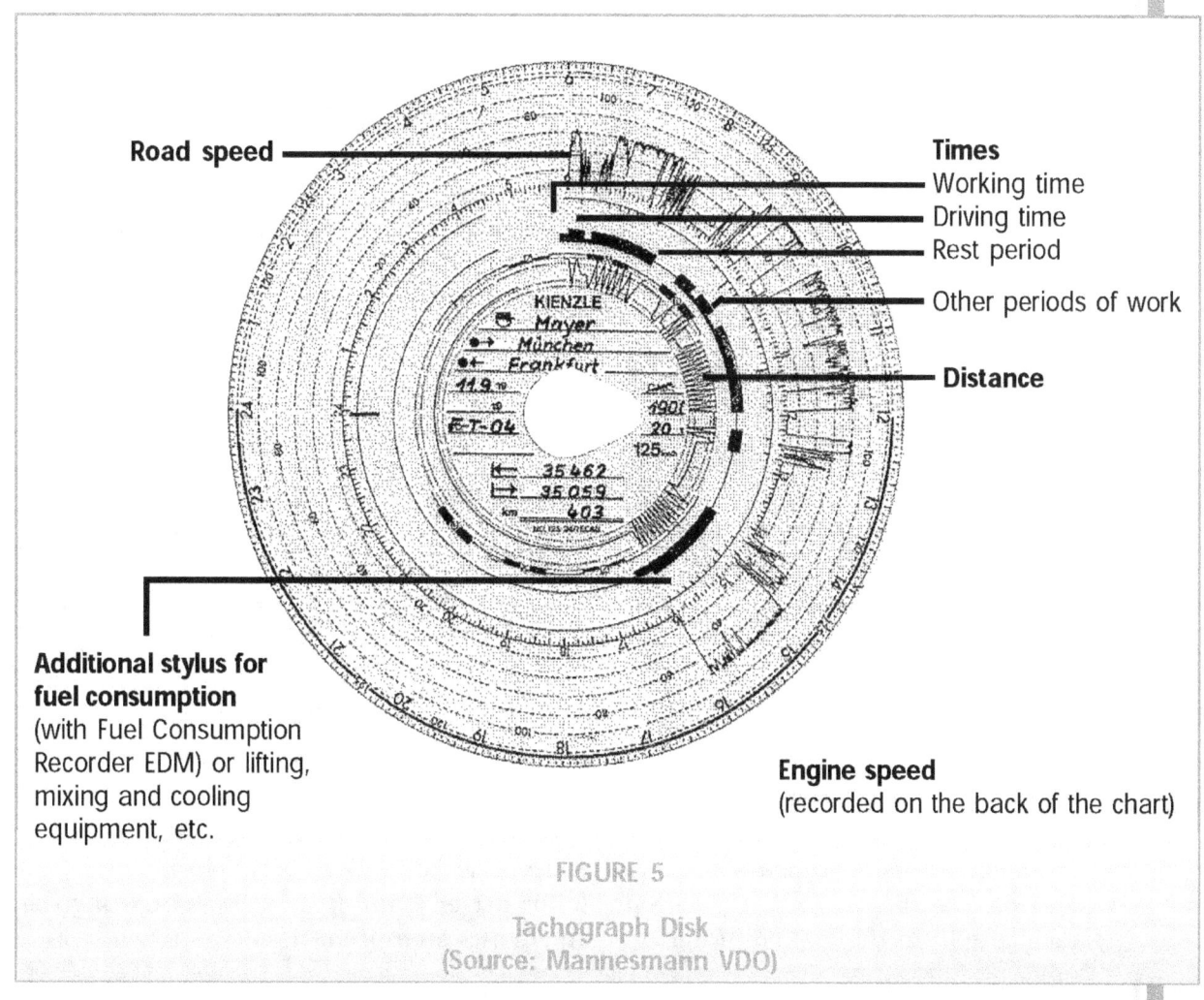

FIGURE 5

Tachograph Disk
(Source: Mannesmann VDO)

To better understand the application of electronic onboard recorders, the Dutch Organization for Applied Scientific Research (TNO) conducted an experiment of black box technology (an onboard data recorder) for the Dutch Postal Service. The black boxes effectively enhanced driver management (better scheduling and increased driving time) and were considered a deterrent against driver violations of driving rules and company policies.

In terms of deploying onboard recorders in the United States and in Europe, data privacy is the most important issue for both companies and drivers. Throughout Europe, protecting individual privacy is considered paramount and laws ensure data privacy with the deployment of onboard recorders. Enforcement personnel can use the recording device only to assess compliance with driving and rest hours. Laws against self-incrimination prohibit the data from being used otherwise.

> ...in the United States and Europe, data privacy is the most important issue for both companies and drivers. Throughout Europe, protecting individual privacy is considered...[and]...Enforcement personnel can use the [onboard] recording device only to assess compliance with driving and rest hours.

A similar attitude affects the use of tachograph results; the data only verify compliance with driving time and rest time regulations. This rule is an integral component of the social regulations. Privacy of data was also referenced in other areas; for example, it is the policy of the Dutch to destroy vehicle-specific weigh-in-motion data after 30 days.

Motor Carrier Approach to Drivers

In both the United States and Europe, commercial drivers are in short supply and time driving is valuable to motor carriers. Labor is the foremost variable in the overall operational equation of the firm. Labor is the single highest cost, and it is subject to the hours-of-service constraints. Deploying tools such as mobile communications and computer-aided routing and dispatching are enhancing the efficiency of the driver and unit.

To maximize driving time, trucking companies would ideally motivate drivers by compensating them for the mile traveled rather than by the hour worked. Driver payment by distance traveled is a common practice in the United States, is the norm in Germany and the Netherlands, but is not allowed in Sweden and France. Sweden's concern is for safety, while France focuses on protecting the driver from abuse. Even in the face of the rules in France (drivers are also paid a minimum 35-hour-per-week salary), French firms offer a bonus as an incentive to their drivers to maximize the number of hours driven.

As in the United States, conscientious motor carriers not only manage their drivers, but they also manage the safety issues – conduct personnel audit documents related to hours of service and provide driver support. Throughout Europe, the *safety director* is responsible for driver training as well as auditing the tachograph. In the United States, focus and costs stem from the driver's hours-of-service logbook, which have generated temporary technology solutions like log-scanners and software. The technology in Europe currently begins with the tachograph and will include the electronic logbook in the future.

Throughout Europe, there is a public and private sector commitment to deploy infrastructure systems to support telematics systems. Germany provides a noteworthy example in its use of global positioning system (GPS) satellites currently in place, new telecommunications capacities, and road operation. Mannesmann has deployed a privately funded traffic detection system in the Bonn area to support its in-vehicle information system; a fleet management solution has been built upon the onboard computer and in-vehicle real-time information platform.

FINDINGS AND RECOMMENDATIONS

The overall European approach is to enhance the value and effectiveness of commercial motor vehicle drivers through rigorous training, education, and performance-based testing and licensing. The traditional regulatory approach is to protect the driver from abusive and unsafe situations and, therefore, to protect their civil liberties.

This European environment yielded a number of key findings and generated recommendations for United States consideration:

- **Comprehensive, standardized driver education curriculum.** A desirable educational program is one endorsed by both public and private stakeholder groups and that advances the skills and business competency of the drivers. Possible approaches include establishing minimum training standards and mandatory training curriculum.

> A desirable educational program is one endorsed by both public and private stakeholder groups and that advances the skills and business competency of the drivers.

- **Performance-based driver assessment.** The goal for training and on-the-road compliance assessment is to ensure maximum driver safety and operational performance. Accordingly, focal points for assessment should be performance-based data that measure driving performance (if necessary, through technology) and outcomes (crashes and violations). These data would improve understanding of driver selection by carriers and create better public policy, such as determining minimum age of driver and hours-of-service rules.

- **Adequate public and private organizations.** Education, testing, and licensing services need to be readily accessible and cost-effective if they are to advance the safety of drivers and their value in the motor freight businesses. The role of private sector advisors holds promise for advancing driver and motor carrier coordination.

CHAPTER 2

- **Human resource management.** In general, a more systematic and scientific approach is needed to manage commercial motor vehicle driver performance. Even in the strict regulatory environment of European drivers, motor carriers provide driver incentives to maximize driving time. It can be demonstrated that an understanding of the firms' objectives can improve profitability for motor carriers.

IMPLEMENTATION STRATEGY

The essence of the deployment strategy is to equip the lead public and private sector stakeholder groups with clear definitions of the benefits of enhanced driver safety and performance. The lead U.S. public agency is the Federal Motor Carrier Safety Administration (FMCSA). There are also many exemplary labor-related safety efforts in the United States – these efforts need to be identified and expanded.

The implementation strategy includes:

- **Involve public and private stakeholder groups.** The FMCSA should highlight the European approach among other government agencies. For example, the Departments of Labor and Education may be interested and able to enhance existing efforts and provide a forum for developing a driver education curriculum, training criteria, and standards. Examples of leading private sector stakeholder groups include truck driver training schools and institutes, insurance company research institutes, and labor unions. Each of these organizations understands the benefits of safety to their respective fields and they share a collective interest in enhanced driver productivity and safety.

- **Build upon successful safety programs.** There are many noteworthy motor carrier safety initiatives and institutions in the United States. For example, the Safety Management Council of the American Trucking Association (ATA) developed the *Safety by Cooperative Partnership Education* (SCOPE) initiative to promote driver education and safety. The ATA Foundation produced *Making A Difference*, a compendium of award-winning safety programs. These initiatives highlight the concept of human resource management and may be the vehicles to broaden its deployment. There are also several exemplary training institutions in the United States. The best practice efforts in the United States can benefit from examining the human resource management approach and from the advances by the lead stakeholder groups.

- **Launch new research efforts to:**

  - Investigate human resource management selection, training media, and driver-performance measures;

  - Determine special resources for establishing driver-training criteria;

- Ascertain possible incentives from insurance companies for efforts favorable to the overall safety objectives and compile the benefits of the leading efforts in the United States.

Equipped with new research findings and the CVS Panel report, lead groups can advance their own self-interests as well as the public good by focusing on issues of driver education and professional competence.

Chapter Three
VEHICLE SAFETY SYSTEMS

The application of information technologies has revolutionized all industries involved in commercial vehicle safety around the world, including truck manufacturers, road operators, and the trucking companies. In Europe, the mechanical rotary tachograph is evolving into an electronic data-capture device, and manufacturers envision it as the platform for an integrated motor carrier operations system of the future. The Europeans have long captured data to enhance their safety systems, and they are developing new vehicle designs, human interfaces, and dynamic new passive and active onboard safety systems.

FOCUS AREAS

The CVS Panel focused on four Vehicle Safety System areas:

- Safety Research Business Models.
- Crash Investigation.
- Vehicle and Safety System Design.
- Onboard Safety.

Safety Research Business Models

Research programs by truck manufacturers and a variety of affiliated and independent research and testing institutes have a long history of enhancing vehicle safety and delivering ever-improving vehicles to meet the demands of the European marketplace. From cab-crashworthiness to dynamic stability, every area of the vehicle, its operations, and its support systems are continuously reengineered and enhanced.

> Research programs by truck manufacturers and a variety of affiliated and independent research and testing institutes have a long history of enhancing vehicle safety and delivering ever-improving vehicles to meet the demands of the European marketplace.

The heavy-duty vehicle manufacturers of Europe are often partners with U.S. firms. Volvo's largest wholly owned subsidiary is in the United States; DaimlerChrysler owns Freightliner; Renault owns Mack Trucks; and PACCAR owns DAF, the Dutch truck manufacturer. Heavy-duty vehicles are manufactured to meet and optimize trucking companies' goods-movement objectives. Although very different size and weight rules exist in Europe and the United States, the most common vehicle configuration for both nations is rapidly becoming the 5-axle tractor-semitrailer.

The essential areas of focus for truck manufacturers are cab-crashworthiness, dynamic stability, and safety systems. Onboard safety systems are divided into passive (reduce the consequences of crashes) and active systems (reduce the number of crashes). Emerging onboard safety applications for trucks include seat belt and under-ride protection; other new technologies include steer-by-wire and electronic braking systems. The following sections detail four different European business models from Sweden, Germany, the Netherlands, and France.

Sweden

Sweden's focus on safety design has resulted in the Swedish cab-crashworthiness test. The test is one to ensure driver survivability. Three impacts – top, front, back – are required and three results are necessary to pass:

1) Survival space is intact.

2) Doors remain closed.

3) Suspension remains intact.

Besides this legal demand, Volvo has an internal, complementary requirement: The doors shall be openable after the test. New systems to increase the driver survivability are being tested and introduced in new models. These systems include soft cab surfaces, such as kneebars, and longitudinal reinforcements in the doors to cope with Volvo's own Barrier Crash Test requirements.

The Swedish system focus on active safety systems is designed to provide drivers with information (collision avoidance and condition information) and adaptation for intelligent speed. Per Adelsson, of Volvo, views the approach to technology as measuring: "The right technology for the right time." Additional new solutions include vehicle multiplex systems, traveler information systems, and airbag electronics. Volvo is advancing its focus from passive to active safety systems that are closely tied to the ability of the driver to accept and use information.

Germany

DaimlerChrysler brings a holistic approach to safety and operational systems. Although regulations primarily address failures in systems, DaimlerChrysler has developed solutions that improve many areas. The company has focused on those technical defects of components responsible for traffic crashes – brakes (60 percent), axles (10 percent), frames (5 to 10 percent), and steering and tires (less than 5 percent) – while advancing an approach to automated control systems. For example, to counter human problems such as the drowsy driver, the overall system includes electronically controlled all-wheel steering systems, intelligent braking, and lane-departure warning systems. DaimlerChrysler also concentrated on the advantages of telematics and the links to commercial vehicle safety systems, including the prospects of an "electronic chauffeur" or long-distance platooning/ automated driving system.

DaimlerChrysler is developing a number of products to address safety as well as operational efficiency, including:

- Telligent Braking® system – reduces braking distance.
- Electronic stability program – uses intelligence to make control interventions.

- Meta-management – delivers information input into systems control.

These initiatives are part of DaimlerChrysler's strategic approach. According to Christian von Glasner, "If one wants to cross the borders of conventional systems in order to improve the active safety of commercial vehicle, one has to use electronic intelligence."

Consistent with its strategic approach, DaimlerChrysler identified the current problem with drowsy drivers. To underscore the extent of the problem, 68 percent of all German commercial vehicle driver-caused fatal crashes result from drowsiness (*Fatal Crashes in Bavaria*, 1991). DaimlerChrysler has developed and tested a prototype lane-departure warning system to address drowsy driving. Using optical lane tracking and an algorithm to detect drowsy behavior, a warning signal produced corrective behavior among test drivers. Test participants reacted favorably to the device and its promise of enhanced safety.

> The DaimlerChrysler model... enhances performance and safety by linking road and vehicle operations systems and information to the driver.

The DaimlerChrysler model includes an approach to telematics designed to accommodate the needs of the driver and create an integrated system that centers on the driver. The model enhances performance and safety by linking road and vehicle operations systems and information to the driver.

The Netherlands

The Organization for Applied Scientific Research (TNO) is a not-for-profit group of 13 institutes. The TNO Traffic and Transport group includes five institutes that focus on infrastructure, road vehicles, applied physics, human factors, and physics and electronics.

TNO investigated a number of high-profile rollover crashes. Crash analysis generated recommendations and a call for a relevant regulation by the EC in Brussels. Because of the rollover analysis, TNO developed a patented tilt monitoring system/rollover warning device, which holds promise for commercial application.

New projects include automated vehicle controls and external speed adaptation. Building on the concept of automated cruise controls (ACC), the Dutch experiment will expand lateral and longitudinal controls, lane tracking, and ACC to use speed control, passing systems, and truck-only lanes. Plans to pilot test an external speed control system include such roadside information data as maximum speed and restrictions. A study is scheduled to explore truck-only lanes.

France

The French government plays a significant role in commercial vehicle research and manufacture. It funds the French National Institute for Transport and Safety Research (INRETS) and owns 49 percent of Renault. The French approach is more decentralized than the other models, yet it is no less comprehensive. The Renault approach is straightforward. Safety is part of productivity and Renault optimizes all areas of vehicle productivity. Further, vehicle safety is part of the overall transportation system and costs are attributed accordingly. The current emphasis is on the active safety research and the development of four systems – braking and longitudinal controls, lateral control, vision enhancement, and drowsiness prevention. An active crash research program influence vehicle systems design.

Crash Investigation

In the United States, the police investigate and report crashes to support adjudication, not necessarily to determine and correct the cause. The litigious environment in the United States has effectively excluded parties other than the police from collecting crash data. In Europe, however, vehicle manufacturers and third-party organizations study crashes to understand causes and take appropriate corrective actions. These analyses support enhanced safety system and vehicle design, and they help frame effective public policy.

> In the United States, the police investigate and report crashes to support adjudication, not necessarily to determine and correct the cause... In Europe, however, vehicle manufacturers and third-party organizations study crashes to understand their cause and take appropriate corrective actions. These analyses support enhanced safety system and vehicle design, and they help frame effective public policy.

The Volvo crash investigation team has been active across Europe since 1969. When possible, the team collects data at crash sites. Analytical results, as well as medical aspects, are key to developing active and passive safety systems. For example, the distribution of injuries by body part (primarily head and chest) shows the need for greater seat belt and airbag (supplementary restraint system) use. Secondary consideration has focused on systems that prevent injury to legs and knees, and corrective improvements are becoming design enhancements. Volvo is also actively developing passenger bunk restraints in truck sleeper compartments. The other two original equipment manufacturers (OEMs) visited (DaimlerChrysler and Renault) also have active crash investigation teams. Throughout Europe, government and third-party organizations also collect and analyze crash information and participate in many areas of manufacture, regulatory compliance, and public policy development.

INRETS, which is funded through the National Vehicle and Road Safety Programme, is also under the control of two French ministries. INRETS created its Transport Safety Epidemiology Laboratory in 1994 to carry out epidemiological studies and in-depth crash investigations. The procedure for analyzing in-depth crash information is emerging from an operational test in four geographic areas of France and in conjunction with French vehicle manufacturers. The approach includes creating a registry of road crash injuries and collecting in-depth data.

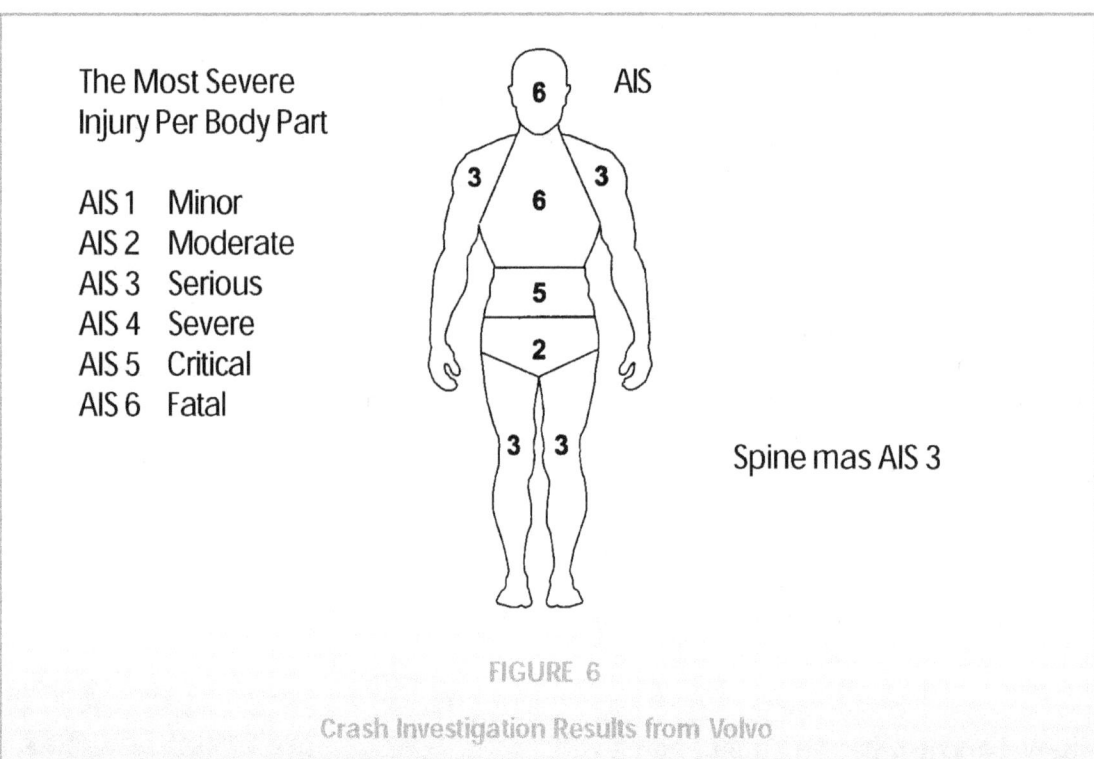

FIGURE 6
Crash Investigation Results from Volvo

The Federal Highway Research Institute (BASt) is funded in part by the German government. BASt is a technical and scientific institute responsible to the Federal Ministry of Transport, Building, and Housing. The Institute acts as scientific advisor to the ministry on technical matters and transport policy. It also plays a leading role in formulating specifications and standards. BASt's International Road Traffic Accident Database is a membership-funded effort that contains information for 30 countries and four continents. In addition to government data, BASt includes special data collected at crash scene investigations.

In conjunction with the National Office for Highway Systems of BASt, the independent DEKRA Accident Research Division has created a database of detailed crash information. DEKRA analysis begins with official traffic crash statistics, but more data were needed to make precise statements. The detailed database, including critical situation information, allowed for potential safety improvements. DEKRA operates more than 80 branches that conduct annual safety inspections and provide fleet management service. The DEKRA Accident Research Division determines crash cause, plays a role in adjudication, and is sometimes summoned to perform a court-ordered role.

BASt, as well as DEKRA and several other institutions in the accident investigation sector, have been heavily involved in evaluating Mannesmann VDO's UDS. Claims from the German Traffic Court conference for introducing mandatory Accident Data Recorder date back to the early '70s. Field trials since then have shown that crashes could be resolved much faster, without making vague assumptions, thus providing for suitable justice. At the same time, preventive effects could be realized, as drivers tend to drive more carefully with the black box onboard. Policy initiatives may ensue as German public authorities, as well as the EC become increasingly aware of this information.

Vehicle and Safety Systems Design

Overall, the safety and operational systems of commercial vehicles are thoroughly designed to optimize performance. According to DaimlerChrysler's Hans-Harald Eggelmann, "We at DaimlerChrysler can arrange for shorter stopping distances of truck-trailer combinations; we can technically control vehicle dynamic behavior under changing road or weather conditions; we can introduce means against tailgating; we can assist and support the driver in his tasks to control the vehicle; we can deliver ambient conditions for the driver to lessen fatigue . . . and as engineers we might propose more. But do existing regulations allow us to introduce into the market what we consider advantageous and safe?"

Mr. Eggelmann pointed out that safety describes conditions in the absence of failure effects. Therefore, safety provisions are set by failure probabilities and the consequent compensation to define object safe conditions – a theme echoed by the Dutch risk-assessment models. This applies to establishing both regulations and appropriate safety standards.

In the United States, the intent of Federal safety rules is to actively promote safety – U.S. code requires instituting preventative standards in order to set motor vehicle safety. The broad approach allows for a variety of systems and technologies that are functionally acceptable. In Europe, however, there is a tradition of national "vehicle type approval" that has extended to the European supranational bodies. Accordingly, there is a tradition of specific standards related to cab-crashworthiness and brake and steering systems. In light of new electronic systems, it is difficult to establish specific standards and determine how the standard is realized.

Standards for Electronic Components

Emerging active safety systems use electronic systems to collect, process, and manage information about the driver, vehicle, and roadway conditions and performance. Although the underlying communications and computing systems are advancing rapidly (functionality expanding and price declining), the lack of standards for interfacing technology threatens the new systems deployment. An essential tenet of active safety systems is the delivery of information to the driver. The link to the driver is critical and is based on information flow and the human-machine interface.

An answer to the standards issue is a standard that will be set through a regulatory rulemaking process of the EC. The DRAFT *Uniform Provisions Concerning the Approval of Vehicles with Regard to the Safety Concept of Vehicular Complex Systems Comprising Electronic Components* has been circulated and a rule is forthcoming.

In support of the effort, the *European Statement of Principles – On Human Machine Interface for In-Vehicle Information and Communications Systems* was submitted in 1998. A broad group of equipment and vehicle manufacturers support

CHAPTER 3

the principles, which address overall design (to neither create a hazard nor distract the driver) as well as specific design elements. These include installation (rules regarding placement, for example, in the line of sight), information presentation (agreed-to symbols, timely information, easily assimilated), interaction with displays and controls (hands-free, logical), system behavior (allow for driving, but take control of the vehicle with failure), and information about the system.

Test and Design Standards

On-the-road crash data are used for design as well as for extensive testing and continuous reengineering. There have long been standards in Europe regarding testing, design, and performance in order to manufacture the safest and most efficient vehicles. Even with these standards, each manufacturer and country has focused on slightly different measures. All are committed to designing safe and efficient vehicles and to ensuring oversight of the design process.

Vehicle Design

The overall approach to design for safety is directed toward creating a more responsive vehicle system and a more intelligent operator. All operating areas of the vehicle are being designed to enhance performance with electronics – for example, electronics applied to brakes, steering, and information flow.

Some of the new areas of active safety systems include:

- Obstacle-detection systems for collision avoidance.
- Drowsy-driver detection and prevention systems.
- Vision enhancement.
- Emergency vehicle control systems.
- Electronic controls for vehicle stability, braking, and steering.
- Load-tilt monitoring and rollover warning devices.

Road Design and Operation

The structure and condition of the roadways are vital to overall roadway safety, and Europe has several noteworthy road design improvements. For example, one new design affects roadway improvements and expansion near Rotterdam. The intent is to reduce the diversity of vehicles in traffic by limiting the number of entrance and exit ramps (excluding local commuter traffic and separating vehicle types) in order to lessen the interruptions to truck traffic flow. Truck-only lanes are being designed and tested as part of the overall "transport in balance" approach of the Dutch.

> ...the Dutch are [designing and testing] truck-only lanes [with the intent to] reduce diversity of vehicles in traffic.

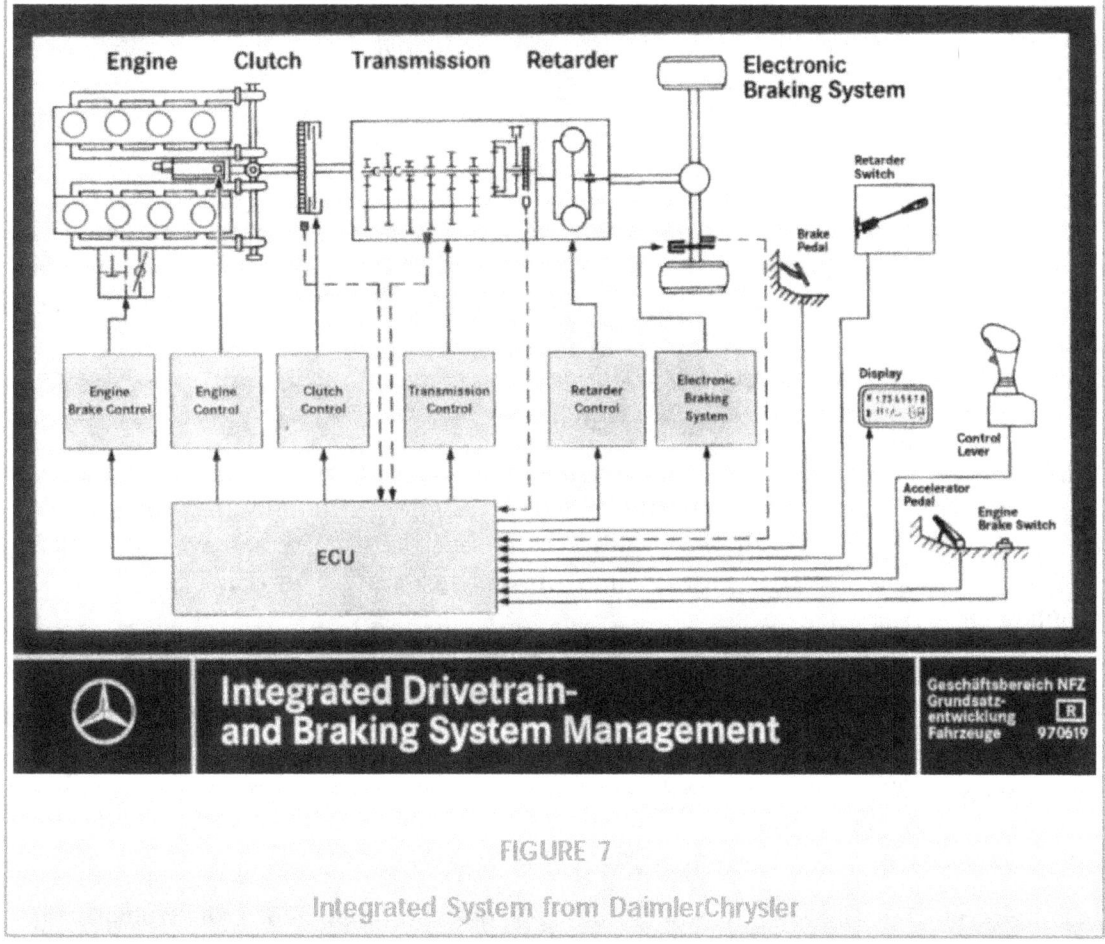

FIGURE 7

Integrated System from DaimlerChrysler

In terms of operations, the Swedes have promoted the use of longer combination vehicles (LCVs) to reduce the number of trucks needed to move the growing volume of freight. This approach is built on the belief that the safety benefit of reducing the numbers of trucks exceeds the potential hazard of LCVs, while increasing roadway safety.

External Safety/Risk Assessment

Assessing risks is part of road policy throughout Europe, particularly in the area of transporting hazardous materials. Risk analysis is used less frequently in the United States, and there is an apparent lack of consistent focus on routing hazardous materials around the country.

Management based on risk assessment and an integrated approach is important in framing political decisions. Comparing risks of hazardous materials transport can only be made when all the safety factors (such as the nature and volume of the transport flow, traffic safety and vulnerability of the surrounding area) are known. Their mutual correlations can be assessed and modes can be compared.

Dutch policy is to decrease the probability of hazardous material incidents, conduct risk management by transportation activities (risk analysis is rational approach), and plan appropriate responses. The Dutch determine risk for transporting

hazardous substances by the two concepts of individual risk and societal risk. They compare the direct consequences for people near transport routes with the potential of serious crash and release of hazardous materials. Germany also uses computer-aided risk analysis of the dangerous goods transport.

Shift of Freight Flow to Rail

There is an aggressive effort throughout Europe to shift freight from truck to rail. The Dutch government has made it policy to shift freight volume to rails and away from road. The Germans demonstrated their commercial vehicle safety problem by pointing out that the truck modal share grew from less than 50 percent in 1970 to 65 percent today. Other European countries also focus on modal shift as a strategy for improving commercial vehicle safety.

Europe has traditionally used the fuel tax to fund social programs and as a means to discourage highway use – regardless, motor freight still dominates. With the failure of the *price* approach, a more direct approach is expected. It will be difficult, however, to change logistics patterns and freight flows built on truck movements to freight rail and other modes.

The efficiency of one small Dutch freight-forwarder, RSC, illustrates the promise of increased rail freight. RSC moves freight from Rotterdam to Germany and points in Eastern Europe. The firm combined innovative yard management with train scheduling to meet the customer needs. Yard management uses of multiunits (four or more oceangoing containers on chassis). The simple scheduling of trains for daily overnight service met the customer demands for timely service.

It is believed that the shift to rail would decrease the number of trucks operating on Europe's roadways and reduce potential conflicts between passenger cars and the trucks. These conflicts generally have more catastrophic results than conflicts between passenger cars.

The speed of European trucks is physically limited, which means that trucks operate at a different range speed than passenger cars. It is believed that a speed differential between cars and trucks creates a safety hazard. Thus, fewer trucks on the road also eliminate the speed differential and improve safety.

Onboard Safety

Basic passive safety systems (those that reduce the impact of a crash) center on the driver. Seat belts and airbags are the most obvious. Although the seat belts are required in heavy-duty trucks, only 10 percent of truck drivers use them, versus 90 percent use in passenger cars. Airbags have been an available option for several years, but few systems have been ordered. The paradox is that the manufacturers and governments aggressively focus on driver safety, yet these basic systems are seldom used.

> The paradox is that [European] manufacturers and governments aggressively focus on driver safety, yet these basic [seat belt and air bag] systems are seldom used.

Active onboard safety systems (those that help prevent crashes) will require new information flow throughout the vehicle and delivering additional information to the driver. There are both institutional and operational challenges to deploying these systems. The institutional difficulties stem from the need to comply with EC regulations and the emerging electronics standards. The operational challenge will be the safe delivery of information to the driver – and the acceptance by the company management.

It appears that the current use of onboard information technology is not as widespread in Europe as in the United States. In Europe, many optimization systems are being tested and the basic communications networks are emerging; these markets and systems are well established in the United States. The deployment of onboard information systems (such as onboard computers, coordinated fleet optimization, and communications systems) is a necessary precursor to more sophisticated safety devices. In order to adopt new systems, drivers need experience with onboard electronics and companies need demonstrated financial benefits. Both drivers and companies must participate if the full potential of the next generation of active safety systems is to be realized. In Europe, the required use of an electronic tachograph in the coming years may increase use of onboard systems.

FINDINGS AND RECOMMENDATIONS

European truck manufacturers are closely tied with manufacturers in the United States, however, the traditional working relationships and organizational structures around them are very different. The overall approach to vehicle and roadway safety design stems from research and crash investigation. Standards have been set to help advance systems and public/private working relationships advance truck design and manufacturer.

Based on the key findings, the CVS Panel recommends the following for U.S. consideration:

- **Third-party organizations.** European third-party organizations play an exemplary role in testing, design, safety management, and new research. The business models provide examples of public and private organizations that share a common safety focus. Creating a greater safety focus for third-party organizations in the United States is desirable. In some cases, crash investigation research and in-company safety audits may require an effort to create new third-party entities to advance the efforts.

- **Crash data.** A large volume of crash information in the United States is collected primarily for litigation purposes. This type of information could

also be used to establish crash causes, which could correct and improve systems operation and vehicle design.

- **Vehicle design standards.** The United States has standards for manufacturing trucks and their systems. The Society of Automotive Engineers, the National Highway Traffic Safety Administration (NHTSA), and others help advance those standards. Europe, however, focuses on testing and licensing standards. Considering standards in the development of passive safety systems (cab-crashworthiness) and active safety systems (electronic interface) holds promise to facilitate new initiatives in the United States. Additional focus should be on making the driver and roadway safer. Current fatigue research in the United States can help frame innovative approaches to drivers and their management. Approaches to enhance driver information management and continuing driver education can help also advance driver and roadway safety. Delivering information and providing support to the driver will depend on the driver's ability to accept and use information.

- **Roadway design and truck access.** Increased congestion means that more passenger cars and trucks must share the road. Risk assessment is a useful procedure for understanding the effect of future roadway design and regulations. Using truck-only lanes in selected locations holds promise for reducing the potential conflict between cars and trucks, particularly during peak travel periods. Innovative new safety systems can be designed, but three nonengineering factors will affect the deployment in onboard applications – government regulations, market acceptance, and product standards. The Europeans continue to struggle with these issues.

European regulations require devices to capture driver data and control speed. This has affected the approach of new safety systems. For other systems, the question is market acceptance. The low airbag and seat belt use demonstrates the lack of acceptance within the market. For the United States, market acceptance of emerging solutions is important. Product standards are also necessary to create a robust market for technology solutions.

IMPLEMENTATION STRATEGY

In Europe, safety is firmly ingrained in the approaches to research, design, and manufacture of trucks. This is also evident in the United States. Accordingly, the lead organizations should be the truck manufacturers and the FMCSA. The groundwork should be set for involving additional third-party organizations.

- **Build upon the efforts of truck manufacturers and encourage third-party organizations.** With the original equipment manufacturers as a focal point, identify other stakeholder organizations for participation. The third-party stakeholders include:

    - Research organizations – university centers, private institutes, insurance company research institutes, and trade association affiliated research groups.

- Other manufacturers: aftermarket parts, peripherals, and electronics.

- Government regulatory agencies and research organizations – State and Federal.

The third-party model in Europe is based on traditional working relationships; in the United States, other relationships are in place as well. The strategy is to encourage these organizations to play larger roles and support performance-based and risk-based analysis during the design phase of commercial vehicle safety.

- **Advance crash investigation and data use.** To advance the use of crash data to enhance safety, the message of how it is useful must be delivered to organizations that can make a difference – organizations that collect the data. The involvement of roadside enforcement officers is important, and a possible lead organization is the International Association of Chiefs of Police. Underlying support may also come from organizations like the National Safety Council (D16 Committee), American Association of State Highway and Transportation Officials (AASHTO), and the Governors' Offices of Highway Safety. Crash investigation procedures for State agencies would also be helpful in capturing valuable data. An organization such as the National Governors Association could create standard crash definitions and classifications. The lead Federal Agencies (NHTSA and the FHWA) should collaborate to address crash causes and promulgate crash investigation methodology. Possible funding should be identified.

- **Encourage use of seat belts and airbags.** The simplest, most readily passive safety devices (seat belts and airbags) are not used in Europe. While passenger car seat belt use is about 90 percent, truck use is only about 10 percent. Airbags, although not fully perfected, are a rarely selected truck option. This paradox should be addressed through promotional activities and public and professional education programs. The leading safety organizations in the United States should maintain a clear focus on market acceptance of emerging safety systems. It is clear that motor carriers will embrace systems that provide clear benefits, but the safety benefits must be clearly identified. As regulatory changes are made, market behavior should be considered to ensure the highest level of participation and success.

Chapter Four
SAFETY REGULATIONS AND ENFORCEMENT

The overall approach of European enforcement is built upon strong rules and deterrence versus strict enforcement. There is a general belief that high compliance exists. Several devices required for the vehicle restrict or monitor performance (for example, tachograph, and speed limiter). Regulations are prescriptive (for example, drivers pay is restricted to per/hour wages in several countries). The burden of regulatory compliance in Europe is shifted to the responsible party, and sometimes shifted to the shipper.

FOCUS AREAS

The CVS Panel focused on three Safety Regulations and Enforcement areas:

- Rules and Regulations.
- Enforcement Model.
- Information Systems.

Rules and Regulations

The European Union (EU) is the confederation of European countries that is moving toward greater economic and organizational integration. The EC was formed as the regulatory body of record and is the arbitrator of final dispute among its member countries. Harmonizing national rules and regulations was a difficult task, and it has resulted in a complicated and complex bureaucratic structure.

The EC promulgates regulations, yet the EC faces continuing challenges as the approaches of once-independent countries merge into a cooperative confederation approach. Each country is obligated to enforce both its own and the EC commercial vehicle safety regulations. As a result, individual country approaches are unique, yet the appropriate European Union and international conventions provide oversight.

Speed

The French and others have identified speed as the foremost safety problem. The problem is so significant that commercial vehicle speed is limited by an engine-governor technology. The 1995 EC Road Traffic Act required a speed limiter that prevents a truck greater than 3,500 kg (7,700 lbs) from exceeding 85 km/h (54 mi/h).

Intelligent speed adaptation (ISA) is an accompanying initiative for reducing speed of passenger cars. The Dutch believe that lowering speeds would reduce hospital admissions by 15 percent and deaths by 21 percent, in addition to providing some positive environmental improvements. ISA technology is evolving; yet the functional requirements are for a variable adjustable speed limiter, signal transmission systems, and vehicle-to-roadside communications.

The Swedes have several ISA tests under way. One test involves focusing a beacon that notifies the driver that the vehicle is exceeding the speed limit. At the opposite end of the spectrum is a totally automatic test. Early results of the automated system that prevents speeding indicate that drivers perceived it as "a safety measure and not an unpleasant control or source of irritation."

Size and Weight

The EC sets size and weight rules. As in the United States, a recent issue is the use of longer combination vehicles. When Sweden and Finland were admitted into the EU, several vehicles were initially banned from operation. They were, however, eventually grandfathered for operation in these Scandinavian countries. Recent discussions have focused on allowing these larger vehicles to operate throughout more of Europe.

In 1997, the EC promulgated a *modular concept* rule for vehicle length in the EU. Operators are allowed to couple their *standard vehicles* (7.8 m/26 ft straight truck, 13.6 m/45 ft semitrailer, and fifth-wheel dolly) in a number of ways to extend the maximum length. This directive allows larger vehicles previously not permitted under earlier EC directive to operate in Sweden and Finland, which restricted maximum length to 16.5 m (54 ft) (semitrailer combination) and 18.75 m (62 ft) (2-unit road-train).

Analysis conducted prior to advancing the modular concept hypothesized the elimination of every third truck trip and deployment of smart logistic solutions. Results showed there was no need for more space in crossing and at roundabouts, and there was a positive effect on traffic safety. The concept is now limited in practice to the two Scandinavian countries – new rules would be needed to allow the concept to expand to other parts of Europe.

TNO, under contract to the Dutch Ministry of Transport, assessed the influence of LCVs. The report focused on traffic process and traffic safety. Analysis considered visibility, vehicle stability, overtaking maneuvers, traffic flow, and crash behavior. The 1996 study found that LCVs should not operate in areas of dense traffic and only certain configurations with specific rules (loading, spacing of brake mechanisms) would be acceptable in The Netherlands.

A number of credible European third-party research institutes and government agencies will debate the issue at the EC level and decide the future of longer combination vehicles and the modular concept.

Enforcement Models

Each European country maintains a commercial vehicle safety enforcement program to ensure compliance with the EU and national rules and regulations. All four countries visited by the CVS Panel presented their programs and each represents a model for consideration.

The Dutch Model

The Dutch enforcement model stems from the belief that prevention is better than the cure. Rising traffic volume has led to the belief that roadside enforcement has

CHAPTER 4

become less viable. Since 1995, the Dutch have developed an innovative approach that combines a warning system and a focused in-company inspection process. The approach stems from a clear understanding of the motor freight industry and concentrates on high-risk carriers.

> ...the Dutch have developed an innovative approach that combines a warning system and a focused in-company inspection process... The roadside inspection activity directly affects the in-company inspection process.

The Dutch model assigns transport operators into three categories:

- Large, well-managed companies that comply with laws as much as possible.
- Medium-sized firms that attempt to comply with laws.
- Firms with inconsistent safety compliance.

Two noteworthy examples of high-risk groups are low-wage foreign carriers that operate unsafely and *re-start* firms that purposely use bankruptcy to avoid creditor and tax obligations. The industry, as well as the regulatory agencies, has agreed to reduce the effect of the high-risk group. Regulatory agencies include:

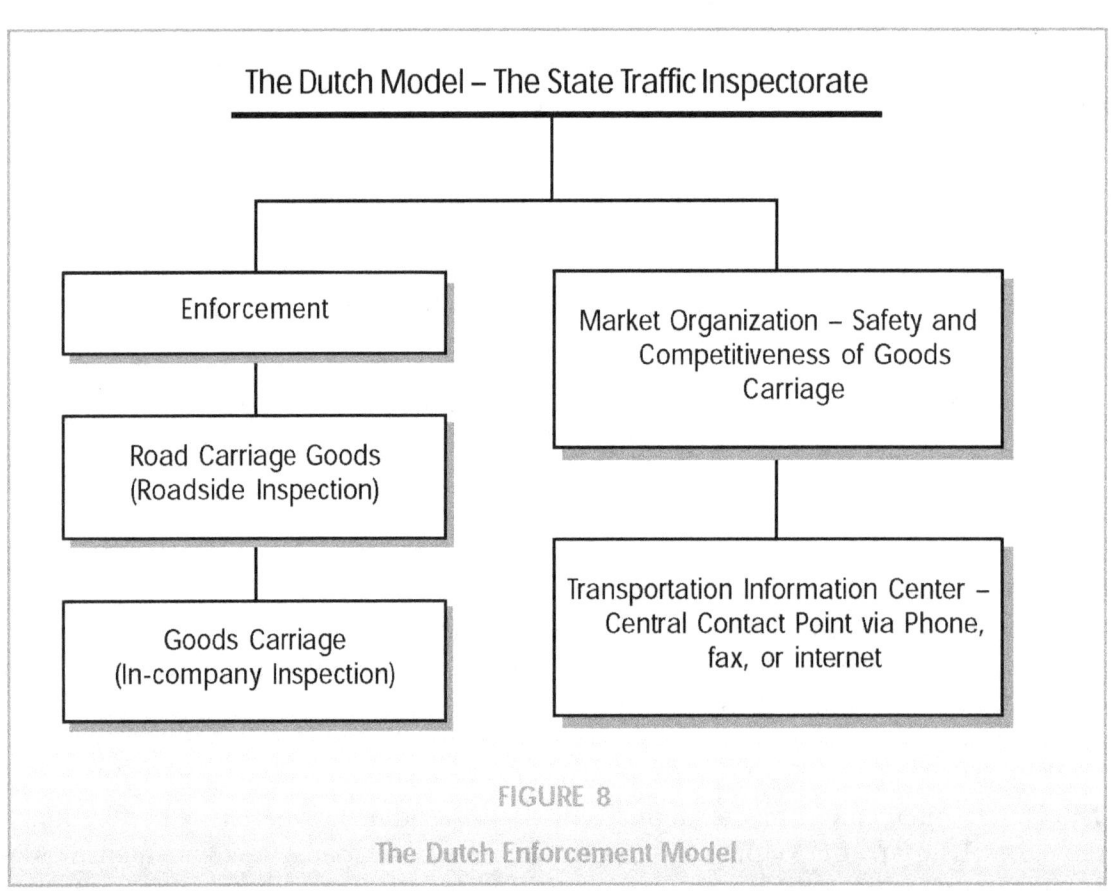

FIGURE 8

The Dutch Enforcement Model

- The State Traffic Inspectorate – checks compliance and disseminates information.
- The Market Organization Division – develops policy and regulations and operates the Transportation Information Center.
- The Enforcement Division – conducts roadside and in-company safety compliance checks.

The roadside inspection activity directly affects the in-company inspection process. The Dutch have conducted a warning system since 1992. A breach of driving or resting hours violation did not generate an official report, rather a written notice was sent to the operator. This process establishes a *danger zone* where steps could be taken to improve compliance. A serious violation produces immediate repressive action. Recurring violations trigger an extensive in-company investigation. The in-company inspections focus on the areas identified in the warning system.

Since 1995, a new roadside checking method has involved two large-scale random checks in April and October. Information compiled creates a two-part approach – market segment investigations and repeat investigations. Compliance improved for several years, but has tapered off. Agricultural haulers (flowers, vegetables, and fruit) and sea-containers are industry market segments identified with higher than average violation rates. These segments have undergone periodic *market segment investigations* in an effort to improve compliance.

The French Model

The French roadside enforcement approach is decentralized into the 22 regions and is conducted by 400 inspectors. The inspectors have three broad, powerful sanctions:

1) Revoke authority.
2) Immobilize vehicles.
3) Decide responsibility – driver, company, or shipper.

The enforcement philosophy is built upon three tenets:

- Ensure fair competition.
- Advance safety on the roads.
- Ensure acceptable work conditions for drivers.

Cooperative efforts between government and industry have shown recent improvements in regulations, and sanctions have encouraged safer motor carrier practices. In 1996, the French Prime Minister moved to enhance the enforcement effort by requiring commercial vehicle safety plans. Additional agency coordination and several new efforts tied with punitive measures were introduced, including authority to search for the gross offenders of driver rest rules, conducting more in-company inspections, and publicity of sanctions against carriers.

CHAPTER 4

The German Model

The German government regulatory ministry for transport is divided into three agencies for taxation and control, statistics and market assessment, and administration. Roadside enforcement activities are notably traditional – 230 controllers operate nationwide with 2-way radios, portable scales, and *sure eye* selection system. The primary causes for truck difficulties are rest-period violations (58 percent), credentials (20 percent), and weight violations (10 percent).

Germany's enforcement practices for drivers' hours-of-rest violation are punitive. There is a demerit system, and if hours of service are exceeded, the driver is assumed to be fatigued and is subject to the relevant criminal code. There is also an insurance incentive for training programs, but it is not tied to a specific carrier's performance. Germany is also experimenting with black boxes that capture vehicle performance data for regulatory and post-crash review.

Germany out-sources annual vehicle inspection activities. DEKRA provides inspection activities and offers other services to 45,000 fleets with 500,000 vehicles. The service extends beyond the required annual inspection and uses data-capture technology to provide maintenance and safety management services, as requested and purchased by the trucking companies. DEKRA also provides financial services and access to its information network and repair facilities.

Uniformity in European Regulations

There is a concern about the differing levels of enforcement between the countries of Europe. France, The Netherlands, Germany, and Sweden each indicated a problem with allegedly unsafe operators from other countries (primarily from the less-developed countries in eastern and southern Europe.)

The EC's eighteenth report, *The Implementation of the Social Regulation Relating to Road Transport* (December 1997), details the regulatory compliance performance for hours-of-service by member countries. There are clearly different approaches by each country, and the EC highlights both successes and failures. The report identifies and discusses charges of discrimination against non-national companies and recommends ways to remedy the situation.

Size and Weight

The Dutch have developed an innovative weigh-in-motion and video system (WIM-VID+) at a test site on the highway between Antwerp and Rotterdam. The system collects data about weight violations and is used to establish strategies for targeting vehicles and reducing overweight violations. The system identifies the magnitude of violations and vehicle types. It identified the most likely gross violators to be 5-axle tractor semitrailer configurations with a 2-axle tractor. To address this group, a 10 percent tolerance was permitted and dramatically increased penalties were put in place for gross violators. The result was favorable. The optical character recognition system proved to be somewhat reliable, but additional functionality is desired.

The most notable part of the WIM-VID+ system was the centralized information flow. The roadside results were monitored by a command center with links to company-specific information and foreign databases. The results of the WIM-VID+ system focused in-company inspection resources, thereby creating a dialogue about regulations, accurate vehicle loading, and change management. The purpose was to improve the safety and weight compliance of the carriers, not to maximize fines or revenue from the system.

Enforcing weight limit requirements is a priority safety objective for the French. Weighing heavy vehicles is believed to enhance safety, maintain competition (within the industry and between the modes), and preserve infrastructure. In response to increasing traffic (and the growing number of overloaded vehicles), the French have developed a strategy of deploying both low-speed and high-speed weigh-in-motion technology to augment the overtaxed static scale system.

> The emerging approach in size and weight enforcement is based on automated data collection and inspection selection system. Action by the enforcement agency occurs either at the roadside or in-company.

The emerging approach in size and weight enforcement is based on automated data collection and inspection selection system. Action by the enforcement agency occurs either at the road-side or in-company. The Germans and French use an approach that combines strict rules with high compliance. Roadside enforcement is based on a *keen eye* approach. When trucks are stopped, checks address only logs, size, and weight. An annual mandatory inspection process considers proper operations of onboard safety systems (such as brakes).

Information Systems

Europeans have not felt the need to use roadside information for screening or credential checks. Accordingly, the need for source data is limited. A centralized system could serve several purposes, including data collection and management for focused enforcement and a method to catch individuals who use repeated bankruptcy to circumvent regulations and business obligations. In general, there appears to be a focus in Europe on complying with the law and less emphasis on litigation, as in the United States. Therefore, there is less need to capture and maintain information.

A number of functions may benefit from information system development, including the following noteworthy areas:

- **Europe-wide commercial driver license information database.** Such a database is planned for deployment in 2000; however, it is expected to be less robust than the Commercial Drivers License Information System in the United States.

- **Inspection results.** Europeans believe that unsafe foreign carriers and varied enforcement levels among EU countries hurt industry's competitiveness. Centralized inspection results may allow analysis of these issues.

CHAPTER 4

- **Decentralized annual safety certification.** An information source could facilitate the annual process and create a data source for random roadside checking.

The deployment of technology-based road management (telematics) is well under way in Europe. There is a desire for intermodal connection and other real-time travel information, but the market for these information systems is still quite limited, and few data sources desired have near-term solutions.

As in the United States, there are several well-established, value-added, and proprietary vehicle information networks. For example, the German DEKRA uses its own communications network to link its facilities and provide services to fleets. The core services are annual safety certification.

FINDINGS AND RECOMMENDATIONS

The Dutch quote, "prevention is better than the cure," captures the essence of the European approach. To support that conclusion, the Europeans have many models of regulations and enforcement systems directed at changing and improving the safety performance of motor carriers. The most noteworthy are the public/private third-party safety consultations. There are examples in the United States of independent certification efforts and effective safety management in firms and in third-party organizations. An expansion of these efforts holds promise for the United States. Based on the European environment, a number of key findings generated recommendations for consideration in the United States:

> ...the Europeans have many models of regulations and enforcement systems directed at changing and improving the safety performance of motor carriers. The most noteworthy are the public/private third-party safety consultations.

- **Alternative inspection activities.** An expanded focus on in-company inspections holds promise to decrease the costs of expensive roadside operations in the United States. Possible alternative models include roadside data to focus in-company inspections, third-party inspectors, and self-certification.

The Dutch model is particularly applicable to the United States because the emerging use of roadside technology, supported by innovative new rules and procedures for in-company selection, is a slight variation on the emerging United States model.

The focus in the United States on in-company inspections with annual inspections by third-party organizations is also noteworthy. There are a number of effective safety management programs within the motor carrier industry and the firms themselves. The annual ATA Safety Management Council's President's Award acknowledges the safety performance and management systems within the leading firms.

In the United States, several third-party safety management firms provide services to motor carriers. These usually focus on carriers at the verge of closure because of poor safety practices. Expanding these types of best practices for safety consulting could also be used for good carriers and act as third-party inspectors for internal compliance reviews.

Unlike in Europe, where the laws hold greater sway, penalties can be established that support a self-certification safety compliance process. Many motor carriers in the United States maintain effective safety management programs. A combination of these programs and enhanced penalties could be the foundation for a self-certification initiative.

Roadside enforcement personnel in Europe are not equipped to access information electronically (carrier ratings, inspection selection system, and driver licenses) – and the agencies do not perceive this as a problem. The exception is the prototype Dutch WIM-VID+ program; it is used to target follow-up in-company inspections and consultation.

- **Consider other European regulations.** European focus areas include speed, size and weight, and modal shift. Speed is directly constrained with a limiter; size and weight is being considered as a method to reduce congestion; and, the Europeans are actively trying to shift freight from truck to rail. Each of these approaches to commercial vehicle safety should be further considered for use in the United States. There appears to be logic in the modular approach for United States *longer combination vehicle* rulemaking. The industry has long gained increases in productivity when safety-neutral outcomes can be documented.

> European focus areas include speed, size and weight, and modal shift. Speed is directly constrained with a limiter; size and weight is being considered as a method to reduce congestion; and, the Europeans are actively trying to shift freight from truck to rail.

IMPLEMENTATION STRATEGY

Many organizations in the United States share the same safety goals, which could be the basis for a new effort to improve safety. The FMCSA is the lead organization for advancing new regulations and enforcement activities models. Given the decentralized nature of the industry and the regulatory enforcement, several industry groups (including the ATA, National Private Truck Council, and Owner-Operators and Independent Drivers Association) and State enforcement or agency groups (for example, the Commercial Vehicle Safety Alliance and AASHTO) should focus on the benefits of alternative models for ensuring safe motor carrier operations. Creating separate bodies within existing professional organizations (such as the Society of Automotive Engineers and ITS America) may also be useful in attracting attention and creating momentum. Changes in the regulations and enforcement models would be best served when proceeded by an all-encompassing policy imperative. As to the specifics, the CVS Panel suggests scheduling a number of different meetings to advance changes in regulations and organizational approaches and to build momentum.

Demonstrating the potential benefits to motor carriers will only go so far, but creating a significant incentive program will encourage the carriers to continue their efforts to improve safety and reduce the number and severity of crashes, thus allowing the government to focus on high-risk carriers. The goal is to create a more compliant motor carrier industry in the United States.

Research efforts in the following four areas could help frame new directions and systems for improving safety:

- Annual safety and truck terminal inspection versus roadside safety inspections.
- Penalty responsiveness.
- Self-certification for safety compliance.
- Third-party roles.

The experience of the current hours-of-service rulemaking process is noteworthy. A large body of assembled research results has given rise to a meaningful public debate and policy development.

Chapter Five

CONCLUSIONS – ADVANCING THE DRIVER, VEHICLE, AND REGULATIONS FOR ENHANCED SAFETY

In Europe, the approach to commercial vehicle safety encompasses three common themes:

- Safety is the absence of failure.
- The driver is the weakest, yet most valuable link in the human-machine interface.
- Integrated systems that maintain clear organizational distinctions and separate identities.

These tenets are the foundation of the European system; they support the rules and regulations of the EU and each country to frame motor carrier operations on Europe's roads. Considered together, these tenets drive the findings and recommendation of this effort.

UNDERSTANDING THE EUROPEAN APPROACH

Safety is the Absence of Failure

As DaimlerChrysler's Hans-Harald Eggelmann clearly stated, this definition of safety resonated through the organizations and institutions that the CVS Panel visited. Quite simply, the philosophy to advancing safety is to eliminate failure.

This approach is clearly demonstrated by using crash analysis (or failures) to design vehicles and performance-based selection of drivers for training as well as employment. The strict rules and regulations regarding safety define the organizational approach to eliminating failure.

The Driver

Rules and regulations, such as hours of service, and well-designed vehicles (as defined by cab-crashworthiness) have traditionally protected the commercial vehicle driver. The paramount concerns of organizational systems have been the safety and civil liberties of the driver (as well as the other users of the roadways).

EC countries, however, face new difficulties given the need to enhance the function of the driver with new information systems and the pressure to enhance driver productivity to meet the needs of business. The Europeans are responding to these demands by developing clear standards that define the man-machine interface, even as they deploy new technologies. While intended to protect the driver, technologies such as the new electronic tachograph can help, but they do not cure all difficulties.

CHAPTER 5

Integration

The EC is the body responsible for new commercial vehicle safety laws. The individual nation's institutions have come together to create a broader community with the EC coordinating rules and regulations.

This synthesis continues as the country entities enforce EU regulations and new issues of coordination arise. The central role of the EC will be further defined by differing levels of enforcement and the concern about regulations and enforcement of less-safe carriers from the less-industrialized members of the EC and Eastern Europe. But enforcement is contingent on the actions of the member countries. Important issues for the future are speed, size and weight, uniformity of enforcement, and creating information sources.

DEPLOYMENT STRATEGY

Based on its findings, the CVS Panel formed recommendations and identified implementation strategies. Panel members believe that these can be the foundation for activities to improve safety by reducing crashes and the number of fatalities. The principal safety efforts of the European countries – *Vision Zero* in Sweden and *Sustainable Safety* in the Netherlands – framed their commercial vehicle safety efforts and drew them together.

Transportation Secretary Slater launched a similar imperative in the United States in June 1999, and this is the catalyst to advance any subsequent efforts. The lead public agencies and private sector organizations should come together to maintain the momentum and launch additional efforts to address specific commercial vehicle issues.

The European Transport Safety Council is a noteworthy safety organization comprised of safety advocates in EC countries, with additional support from the Dutch and Swedish road administrations. Its charge is to advance road safety management through education and promotion. A national safety conference could be held at an appropriate time to help advance safety in the United States.

There are many exemplary programs in both the United States and Europe. The key is to identify the lead organizations in the United States and persuade them to champion the next steps in advancing commercial vehicle safety. The lead organizations should include the Federal Agencies charged with commercial vehicle safety, the truck manufacturers, driver trade associations, motor carriers and training institutions, and insurance companies. The best practices should emphasize the strengths of existing efforts in the United States and be reinforced by the European findings.

Appendix A
AMPLIFYING QUESTIONS

INTRODUCTION/BACKGROUND

- What is the public perception of truck safety, truck drivers, truck size and weight, and the motor freight industry?
- What is the institutional relationship between industry and regulatory agencies?
- What is the level of technological sophistication of the motor carriers, regulatory enforcement agencies, and roadway operators?
- What institutions are driving regulatory change and technology advancement? And why?

HUMAN FACTORS

- How is truck driver performance measured? What are the criteria, calculation of rating, enforcement, etc.?
- What are the regulations related to driver's fitness-for-duty – fatigue, drugs, alcohol, etc.? What are the methods to measure and track? What are enforcement methods and punitive approaches? What are alternative compliance activities?
- Are there Share-the-Road programs (motorist and truck driver education, outreach)? Are there other programs?
- How are drivers trained? What comprises training programs?
- What training technologies are deployed or being tested? What are desired?
- Are in-vehicle collision avoidance or driver-alert technologies being deployed or developed?
- Are roadside alertness devices (rumble strips, raised lane markers, etc.) used?
- Are alertness/fatigue-measuring devices being tested or deployed?

EQUIPMENT AND INFRASTRUCTURE OPERATIONS

- How is vehicle safety measured? What are the industry operational/maintenance activities, regulatory requirements, and enforcement activities?
- Is there uniformity in vehicle safety requirements?
- Are there inspection standards? What is the standard-setting organization?
- What are the roadside devices used for screening vehicles, inspection, data-capture, and communication?
- What are the in-vehicle devices used for monitoring safety-load or brake alignment sensors, etc.?

APPENDIX A

- How prevalent are remote sensing activities like real-time performance monitoring by dispatcher?
- Are operational information services (real-time traffic, rest-stop location and space availability, etc.) provided to motor carriers? What are they?
- Are there any incentives or disincentives for vehicle safety? For example, value-based property tax may encourage use of older vehicles.
- How is truck performance or crash data used for planning?

INSTITUTIONS

- What is the structure of the motor carrier industry – role of motor freight, industry segmentation, economic regulation, and operational objective/performance measurement?
- How are the industry associations or organizing bodies assembled – voluntary trade associations, rate bureaus, etc.?
- What is the structure of the enforcement community – organizational connection, operational philosophy and overall approach? How are they funded?
- What are other regulatory agencies?
- What is the role of union in all institutions – shippers, drivers and governments?
- How does information flow? Computer and communications and non-technological flows-between institutions (agency to agency, industry to agency, etc.)
- What other institutions are involved – universities, other modes, etc.?

Appendix B
COMPOSITE AGENDA

Stora Holm–Monday 9/14/98–Host: Margareta Morck, SNRC
    Introduction–Anders Linquist
    Education program–Tommy Emanuelsson
    Simulator/interactive training
    Traffic safety–Per Dahl
    Commercial vehicle traffic–Margareta Morck
    Road Infomatics and Traffic Management–Torbjorn Bidding/Chris Patten
    Driving license procedures–Leif Hogstrom
    Safety from a truck company–Bengt Gustavsson

Volvo–Tuesday 9/15/98–Host: Lars-Goran Lowenadler, Volvo
    Vehicle dynamics–John Aurell
    Crash research–Lennart Svenson, et al
    Collision safety–Anna Mattsdotter
    Crash analysis–Mario Ligoic
    Driver Environment–Staffan Wendoborg
    Look at the new FM Truck
    Vehicle electronics–Per Adelssonn
    Driver Information Systems–Bo Lind
    Regulations–Bengt Thompsson, et al

Daimler Benz–Wednesday 9/16/98–Host: Dr. Breitschwerdt, DaimlerChrysler
    New developments in European regulations–Eggelman
    Active and passive safety research and development–Wolf and Dr. Pflug
    Potential of telematics–Dr. Ball
    Route guidance, safety, and information technology–Schussler
    Automatic "chauffeur driven" operations–Schultze
    Lane departure warning system–Mrs. Mehring
    Demonstrations on the test track

BASt-Technical–Thursday 9/17/98–Host: Dr. med Bernd Friedel
    Mannesmann-Schmidt-Cotta
    DEKRA–Niewohner (crash investigation) and Kuhlmann (crashes)
    Mannesmann autocomm–Scafer (ATIS and fleet management)
    BAG-Maiworm (roadside enforcement)

BASt-Human Factors–Friday 9/18/98–Host: Prof. Dr. Gunter Kroj
    IRTAD–Berns (international data) and Elsner (national data)
    Traffic medicine–JOO (medical requirements)
    Safety concepts–Hundhausen (driver training and licensing)
    ASF-Korn (simulator)

Ministry of Transport, The Hague and Delft–Monday 9/21/98–Host: Dr. Attema
    Sustainable Safety–Elsenaar
    Structure of Industry–Kastelijn
    Transport Safety Policies–Doornink
    External Safety Policy–v/d Brand
    Introduction to TNO and Roll-over–Hoogvelt
    Passive Safety–de Coo
    Predicting and Preventing drowsiness crashes–Gobel
    Automated vehicle guidance–van Arem

Ministry of Transport, The Hague and Rotterdam–Tuesday 9/22/98–
Host: Dr. Attema
- RVI (Government Traffic Inspection)–Aarsen, et al
- Demonstration of WIMVID
- Transport in Balance–Uitenbogaart
- Waalhaven Terminal–Hoenders

Ministry of Public Works, Paris–Wednesday 9/23/98–Host: Jean Guillot
- Crash Data–Philippe Groleau
- Training of Drivers–Mrs. Daillet-Demets
- Organization of control–Jeanne-Marie Sabattier
- Social regulation–the tachograph-Mrs. Giraud
- THOMSON–Rene Jaouen
- Driver training–Jean-Claude Claverie
- INRETS–MM. Medevielle and Hamelin

BSA International (Chilly Mazarin) and Renault (Guyancourt)–Wednesday 9/24/98–
Hosts: Mr. Vogt (French DOT) and Bernard Farve (Renault)
- Haulage Company–BSA International
- Crashology–Patrick Botto
- Passive Safety–Bernard Favre
- Driver–Patrick Botto
- Active Safety–Bernard Favre

AFT-IFTIM, Minchy-Saint-Elosi–Friday 9/25/98–Host: Jacques Hervo

Jean-Pierre Liano, Director of Development and External Affairs, AFT-IFTIM

Appendix C
CVS Panel Biographical Information

Katherine Hartman, CVS Panel Chairperson, is a transportation specialist with the U.S. Department of Transportation's (U.S. DOT's) Federal Motor Carrier Safety Administration (FMCSA) in Washington, DC. Assigned to FMCSA's Office of Technology Evaluation and Deployment – Technology Division, Ms. Hartman is currently the Platform Technical Director for Commercial Vehicles on the U.S. DOT's Intelligent Vehicle Initiative (IVI). The IVI is a Department-wide initiative – together with the motor vehicle and trucking industry, State and local DOTs, and other stakeholders – to accelerate the development, introduction, and commercialization of driver assistance products to reduce motor vehicle crashes. Ms. Hartman is a graduate of the University of Virginia and holds an M.B.A. from the University of South Dakota.

Bob Pritchard, CVS Panel's Report Facilitator, is a senior associate at Cambridge Systematics (CS) in Cambridge, MA. Mr. Pritchard is a specialist in applying information technologies in motor carrier operations and is developing CS's commercial vehicle operations practice. Mr. Pritchard joined CS in 1998. Prior to that, and since 1989, he served as executive director of the American Trucking Association's (ATA's) Northeast Transportation Institute (NTI). At ATA, Mr. Pritchard's efforts focused on intelligent transportation systems in commercial vehicle operations (ITS/CVO), strategic planning and institutional analysis for ITS/CVO deployment, and the costs and benefits of newly emerging technologies. In June 1997, Mr. Pritchard launched FleetForward, an operational test of the delivery of real-time travel information in motor carrier routing and dispatching. He is a member of the Transportation Research Board's Urban Goods Movement Committee and is chairman of ITS America CVO Outreach Committee. An economist, Mr. Pritchard holds master's and bachelor's degrees from Boston College.

Ken Jennings directs the Truck Size and Weight Program for the Virginia Department of Transportation (VDOT) where he is responsible the statewide program of truck weighing operations. In addition, Mr. Jennings represents VDOT on many projects, task forces, and organizations concerned with commercial vehicle operations (CVO) and intermodal freight issues. These include the Virginia Trucking Advisory Committee, deputy project manager of Virginia's Commercial Vehicle Information Systems and Networks (CVISN) Prototype Project, ITS America (CVO Outreach, ITS Electronic Payment Services Task Force, and the CVO Architecture and Standards Subcommittee), the CVO Working Group of the I-95 Corridor Coalition, chairman of the Southeastern States Inter-Regional CVO Institutional Issues Work Group Steering Committee, Great Lakes States Regional Mainstreaming, and various other ITS/CVO-related initiatives. Before joining VDOT, Mr. Jennings spent more than 16 years in the private sector where he was involved in several multidisciplinary areas of the scale industry. He developed particular expertise in weighing technology, process design, and information transfer.

APPENDIX C

Jim Johnston is President of the Owner-Operator Independent Drivers Association (OOIDA). He has served in this capacity for the past 23 years following 2 years as executive vice president of OOIDA. Under Mr. Johnston's leadership, OOIDA has evolved into the largest owner-operator association in the trucking industry. Mr. Johnston has served on numerous research panels of the National Academy of Science Transportation Research Board, the Congressional Office of Technology Assessment, U.S. General Accounting Office, and others dealing with various aspects of trucking operations such as commercial vehicle safety and truck size and weight issues. Mr. Johnston currently serves on the National Motor Carrier Advisory Committee to the U.S. Department of Transportation. He is a member of the Commercial Vehicle Safety Alliance Senior Strategic Advisory Committee and the ITS America Commercial Vehicle Operations Policy Subcommittee, for which he chairs the committee's Data Privacy and Control Task Force. An outspoken advocate for the rights and well-being of professional truckers, Mr. Johnston is concerned with all areas of commercial vehicle safety.

Ron Knipling is Chief of the Research Division of the FMCSA's Office of Technology Evaluation and Deployment. In this capacity, he manages and coordinates FMCSA's diverse Research and Technology (R&T) program, including research on human factors, technology applications, information analysis, and regulatory reform. FMCSA's human factors program focuses on the problem of commercial driver drowsiness/fatigue and includes nearly 20 different R&T and outreach projects relating to this issue. Many of these R&T projects relate directly to current FMCSA rulemaking relating to commercial motor vehicle driver hours-of-service. Dr. Knipling, who earned a doctorate in physiological psychology from the University of Maryland, most recently served for 6 years as an engineering research psychologist with the National Highway Traffic Safety Administration (NHTSA). At NHTSA, he managed research projects relating to in-vehicle drowsy driver detection, statistical analysis of the drowsy driver crashes, crash problem size analysis, crash causation analysis, and Intelligent Transportation System crash avoidance systems. Dr. Knipling's career includes more than 20 years' experience in behavioral, human factors, and traffic safety research. He is the author of numerous publications and technical reports.

C. John MacGowan currently serves as the Acting Director for the FMCSA's Office of Bus and Truck Standards and Operations. Mr. MacGowan was previously the chief of the Intelligent Systems and Technology Division of the Office of Safety and Traffic Operations R&D of the FHWA. Since 1970, he has held a number of positions in the highway transportation field. Following 10 years in research, where he pioneered computerized traffic signal control systems, Mr. MacGowan spent 5 years in NHTSA, where he served first as the cognizant official for the Fatal Accident Reporting System, and then as the Chief of the Information Management Division where he was responsible for all fatal and special investigation crash data. In 1985, he returned to the FHWA's Office of Motor Carriers as the Chief of the Motor Carrier Information Division where he oversaw innovative work in managing safety information files. He also was responsible for implementing of many truck size and weight regulations. Following this, he spent 1 year as a White House Fellow with a major insurance company and served an additional year as the special assistant for university affairs to the Federal

Highway Administrator. Mr. MacGowan holds a bachelor's degree in civil engineering and a master's in Transportation Engineering.

Larry Oliphant is an independent transportation consultant. He is the current president of the Western Highway Institute and serves on the board of directors of the National Private Truck Council, the Professional Fleet Management Institute, and the Truckload Carriers Association. He is also a consultant to the Truck Renting and Leasing Association. In his consulting work and above directorships, Mr. Oliphant is involved and in all areas of vehicle safety as well as current and future regulatory, policy, and legislative issues. Mr. Oliphant's 37-year career has included positions as vice president, Parts, for International Harvester Corp; vice president for sales and marketing, Volvo White Truck Corp.; and president, Volvo GM Trucks of Canada. Mr. Oliphant holds a bachelor's degree from Northwestern University.

Mike Onder is a program manager for Intelligent Transportation Systems in the FHWA. Mr. Onder is currently responsible for the ITS program for commercial vehicle operations, intermodal freight, and highway-rail crossings. He is leading the effort to develop a construct for using ITS technology in the intermodal freight arena. Prior to joining the staff of the ITS Office, he served as research director for the Florida Division of Motor Vehicles. He also served as the assistant staff director for transportation with the Florida Legislature's transportation committee, and as deputy executive director for the American Association of Motor Vehicle Administrators (AAMVA) in Washington, DC. With the AAMVA, he helped to pioneer the International Registration Plan in the United States and Canada – a system to prorate motor carrier registration fees based upon travel distance within a registering jurisdiction. In private industry, Mr. Onder served as a consultant and sales manager for Amtech Systems Corporation, a manufacturer of radio frequency identification systems. Mr. Onder is a graduate of Florida State University with a bachelor's degree in Political Science and Economics and a master's degree in Business and Public Administration.

Charles Sanft is currently assigned to the Office of Investment Management, Minnesota Department of Transportation. This assignment includes, among other duties, managing Minnesota's involvement in the Interstate 35 Trade Corridor Study. Mr. Sanft has nearly 30 years of transportation experience in the public sector ranging from travel modeling to freight policy analysis, planning, and program management and project development. From 1982 to 1989, he was director of Rail Planning and Program Development, in charge of State and Federal rail programs and projects. From 1989 through 1993, he directed Truck & Economic Studies for the Minnesota Department of Transportation. Since 1993, he has been involved in developing freight transportation functions and organization for the Department as the Freight Policy Director and as Director of Freight Planning and Development. Mr. Sanft is a member of the AASHTO Subcommittee on Highway Transport and chairs that Subcommittee's Truck Size and Weight Task Force. He holds a bachelor's degree in Geography from the University of Minnesota.

Appendix D

CONTACTS/WEB ADDRESSES OF PARTICIPANTS AND ORGANIZATIONS

EUROPEAN CONTACTS: COMMERCIAL VEHICLE SAFETY SCANNING REVIEW

Sweden

Margareta Mörck
 Public Transport and Commercial
  Traffic Division
 Swedish National Road
  Administration
 S-781 87 Borlänge
 +46 243 756 04
 +46 243 756 30 fax
 margareta.moerck@vv.se

Torbjörn Biding
 ARENA Project Manager
 Swedish National Road
  Administration
 S-405 33 Göteborg
 +46 31 63 50 00
 +46 31 15 56 24 fax
 torbjorn.biding@vv.se

Tommy Emanuelsson
 Utbildningscentrum Stora Holm
 417 46 Göteborg
 +46 31 705 68 01
 +46 31-705 68 23 fax
 ushtoe@storaholm.educ.goteborg.se

Bengt Gustafsson
 Kurt Jonssons Åkeri, AB
 S. Hildedalsgatan 13
 SE-402 76 Göteborg
 +46 31 51 35 30
 +46 31 23 95 95 fax

Ragner Fast, Vise President
 Volvo Truck Corporation
 Dept. 23050, VLH8
 SE-405 08 Göteborg
 +46 31 66 47 47
 +46 31 66 20 10 fax
 vtc.ragnart@memp.volvo.se

Lars-Goran Löwenadler
 Volvo Truck Corporation
 SE-405 08 Göteborg
 +46 31 765 15 21
 +46 31 66 66 20 fax

W.S. (Skip)Yaekel
 Volvo Trucks, North America
 7900 National Service Road
 Greensboro, NC 27409
 336/ 393-2825
 336/ 393-3000 fax
 vtna.yaekel@memo.volvo.com

Germany

Reinhard Ball
 Transport Policy and Strategy
 Daimler-Benz AG
 HPC F 607
 D-70546 Stuttgart
 +49-711-17-2 02 65
 +49-711-17-5 31 84 fax

Hans Christian Pflug
 Commerical Vehicles Division
 Mercedes-Benz AG
 D 209
 D-70322 Stuttgart
 +49 -711-17-5 52 14
 +49-711-17-5 36 95 fax

Hans-Harald Eggelmann
 VP Industrial Relations
 Daimler-Benz AG
 HPC B 303
 D- 70546 Stuttgart
 +49-711-17-2 26 74
 +49-711-17-5 21 91 fax

Gerhardt Hauschultz
 Daimler-Benz AG
 D-70322 Stuttgart, Germany

Bill Gause
Freightliner Corp.
P.O. Box 3849
Portland, OR 97206-3849
503/ 735-7413
503/ 735-6800 fax

Andreas Kuhlman
Dekra Automobil AG
Rungestraße 9-10
D-25437 Neumünster
+49 43 21 90 75 0
+49 43 21 542 34 fax

Claus Korn
Simulation Systems Division
STN Atlas Elektronik, GmbH
Sebaldsbrücker Heerstr., 235
D-28305 Bremen
+49 421 4 57 28 51
+49 421 4 57 38 14 fax
ckorn@stn-atlas.de

Bernd Klott
ASF GmbH
Senator-Harmssen Straße, 3
D-28197 Bremen, Germany
+49 421 520 18 24
+49 421 520 17 50 fax
asfgmbh@t-online.de

Ralf-Roland Schmidt-Cotta
Mannesmann VDO, AG
D-78006 VS-Villingen
+49 7721 67 33 32
+49 7721 67 22 75 fax

Hubert Schäfer
Mannesmann Autocom GmbH
Niederkasseler Lohweg, 20
D-40547 Düsseldorf, Germany
+49 02 11 53 68 4 15
+49 02 11 53 68 5 04 fax
passofleet@mac.de

Gunter Zimmerman
International Cooperation
BASt
Brüderstraße, 35
D-5060 Bergisch-Gladbach
+49 220 44 32 61
+49 220 44 39 73 fax

Günter Kroj
BASt
Brüderstraße, 53
D-5060 Bergisch-Gladbach
+49 22 04 43 0
+49 22 04 43 8 33 fax

Bernd Friedel
Automotive Engineering Division
BASt
Brüderstraße, 53
D-5060 Bergisch-Gladbach

Robert Maiworm
Bundesamt für Güterverkehr
Werderstraße, 34
D-50672 Köln
+49 221 5776-120
+49 221 5776-104 fax

Netherlands

Th. W.H.J. Aarsen, Director
Freight Transport Directorate
Box 2501 HS, The Hague
+31 70 30 52 800
+31 70 30 52 777 fax

Peter M.W. Elsenaar, Director
Traffic Safety and Vehicle
    Directorate
Ministry of Transport, Public Works
    and Water Management
Box 20901
2500 EX, The Hague
+31 70 351 67 69
+31 70 351 64 17 fax
peter.elsenaar@dgp.minvenw.nl

G.H.Doornink
Safety Management Division
Ministry of Transport, Public Works
    and Water Management
Box 20904
2500 EX, The Hague
+31 70 351 15 25
+31 70 351 15 98 fax
gert.doornink@dgg.minvenw.nl

Jack M. Van Nieuwenhoven
　Policy Advisor, Safety Management
　　Division
　Ministry of Transport, Public Works
　　and Water Management
　Box 20904
　2500 EX, The Hague
　+31 70 351 15 34
　+31 70 351 15 98 fax
　jack.vnieuwenhoven@dgg.minvenw.nl

M.G. Koopmans
　Senior Policy Advisor, Traffic
　　Management Division
　Ministry of Transport, Public Works
　　and Water Management
　Box 20904
　2500 EX, The Hague
　+31 70 351 15 62
　+31 70 351 15 48 fax

G.A.M. Schipper
　Senior Policy Advisor, Traffic
　　Management Division
　Ministry of Transport, Public Works
　　and Water Management
　Box10700
　2501 HS, The Hague
　+31 70 30 52 716
　+31 70 30 52 777 fax
　gschipper@rvi.minvenw.nl

Hanno (J.E.) Uitenboogaart
　Coordinator, Transport in Balance
　Box 20904
　2500 EX, The Hague
　+31 70 351 14 35
　+31 70 351 14 78 fax

Gerard J.M Meekel
　RDW, Vehicle Technology and
　　Information Centre
　Box 777
　2700 AT, Zoetermeer
　+31 79 34 58 334
　+31 79 34 58 041 fax

G.R.M. Jansen
　Managing Director
　TNO Traffic and Transport
　Box 6033
　2600 JA, Delft
　+31 15 269 68 70
　+31 15 269 77 82 fax
　g.jansen@inro.tno.nl

P.J.A. de Coo
　Senior Engineer, Crash Safety
　　Research
　TNO Road Vehicles Research
　　Institute
　Box 6033
　2600 JA, Delft
　+31 15 269 63 50
　+31 15 262 43 21 fax
　Coo@wt.tno.nl

R.B.J. (Boudewijn) Hoogvelt
　Vehicle Dynamics Department
　TNO Road-Vehicles Research
　　Institute
　Box 6033
　26oo JA , Delft
　+31 15 269 64 11
　+31 15 269 73 14 fax
　Hoogvelt@wt.tno.nl

France

Catherine Marque
　International Affairs
　Sécurité Routière
　Arche de la Défense
　92055 La Défense Cedex, Paris
　+33 1 40 81 80 73
　+33 1 40 81 81 71 fax

Elisabeth Pillet
　International Affairs Bureau
　Ministry of Public Works,
　　Transportation and Housing
　Arche de la Défense
　92055 La Défense Cedex, Paris
　+33 1 40 81 87 35
　+33 1 40 81 17 22 fax
　pillet@dtt.equipement.gouv.fr

Phillipe Groleau
National Observatory for Road Safety
Sécurité Routière
Arche de la Défense
92055 La Défense Cedex, Paris
+33 1 40 81 80 28
+33 1 40 81 80 99 fax

Jean Guillot
Deputy Director for Surface Transportation
Sécurité Routière
Arche de la Défense
92055 La Défense Cedex, Paris

Jean-Pierre Médevielle
Adjoint Director General
INRETS
Centre de Lyon-Bron
25 ave. François Mitterand, Case 24
F-69675 Bron Cedex
+33 4 72 14 23 40
+33 4 72 37 84 24 fax
jean-pierre.medevielle@inrets.fr

René Jaouen
Project Manager, Civil Application Department
Thomson-CSF
10, ave de la Ière D.F.L.
29283 Brest Cedex
+33 2 98 31 23 08
+33 2 98 31 27 36
rene.y.j.jaouen@rcm.thomson.fr

Alain Flipo
Thomson Training & Simulation
48073 Saint Honore
+33 1 34 90 35 08
+33 1 34 90 35 43 fax

Rémy Boinot, President
BSA International
30-32, route de Longjumeau
91385 Chilly-Mazarin Cedex
+33 1 69 10 17 10
+33 1 69 34 32 59 fax

Guy Gallo
Export Manager
BSA International
30-32, route de Longjumeau
91385 Chilly- Mazarin Cedex
+33 1 69 10 17 10
+33 1 69 34 32 59 fax
bsagallo@francemultimedia.fr

Bernard Favre
Head of Advanced Engineering
Renault V.I.
1, ave. Henri Germain
69802 Saint-Priest Cedex
+33 4 72 96 45 84
+33 4 72 96 61 89 fax
bernard.favre@renaultvi.com

Andras Kemeny
Research Group Manager
Renault
1 ave du Golf, F-78200 Guyancourt Cedex
+33 1 34 35 19 05
+33 1 34 95 27 30 fax
andras.kemeny@renault.fr

Jean-Marc Kelada
Research Engineer, Driving Simulators
Renault
1 ave du Golf, F-78200 Guyancourt Cedex
+33 1 34 95 19 67
+33 1 34 95 27 30 fax
jm.kelada@renault.fr

P. Botto
European Center for Security and Risk Analysis Studies
Hôpital Nord -- Accidentologie
SEFAL 2
Place Victor Pauchet
80054 Amiens Cedex 1
+33 3 22 66 83 54
+33 3 22 66 86 24 fax

Jean-Pierre Liano
Director of Development
AFT-IFTIM
60290 Monchy-Saint Eloi
+33 1 42 12 51 78
+33 1 44 66 37 90 fax
liano@aft-iftim.asso.fr

APPENDIX D

Jacques Hervo
   Manager
   AFT-IFTIM
   60290 Monchy-Saint Eloi
   +33 1 44 66 37 49
   +33 1 44 66 37 90 fax

Dominic Chaumet
   Technology Education (Pilote 2001)
   AFT-IFTIM
   60290 Monchy-Saint Eloi
   +33 1 44 66 37 16
   +33 1 44 66 37 45 fax
   chaumet@aft-iftim.asso.fr

Francois Serrier
   Technology Education
   AFT-IFTIM
   60290 Monchy-Saint Eloi
   +33 1 44 66 38 97
   +33 1 44 74 06 69 fax
   serrier@aft-iftim.asso.fr

Jacques-Claude Rennesson
   Intelligent Transportation Systems
      Unit
   AFT-IFTIM
   60290 Monchy-Saint Eloi
   +33 1 44 66 37 92
   +33 1 44 66 37 60 fax
   rennesso@aft-iftim.asso.fr

ORGANIZATION WEB SITES

France

AFT-IFTIM
www.aft-iftim.asso.fr

French National Institute for Transport and Safety Research (INRETS)
www.inrets.fr

Thompson
www.thomson-europe.com

Thompson Training and Simulation
www.tts.thomson-csf.com

Germany

BASt's International Road Traffic Crash Database
www.bast.de/irtad

Mannesman VDO
www.passo.de

Sweden

Stora Holm Training Center
http://transport.storaholm.educ.goteborg.se

The Netherlands

Applied Scientific Research (TNO)
www.tno.nl

United States

American Trucking Associations
www.trucking.org

Cambridge Systematics
www.camsys.com

Federal Highway Administration
www.fhwa.dot.gov

FHWA Office of International Programs
www.international.fhwa.dot.gov

Federal Motor Carrier Safety Administration
www.fmcsa.dot.gov

Minnesota Department of Transportation
http://www.dot.state.mn.us

National Highway Traffic Safety Administration
www.nhtsa.dot.gov

Owner-Operator Independent Drivers Association
www.ooida.com

Virginia Department of Transportation
www.vdot.state.va.us

FHWA INTERNATIONAL TECHNOLOGY EXCHANGE REPORTS

Infrastructure

Geotechnical Engineering Practices in Canada and Europe 🖰
Geotechnology—Soil Nailing 🖰
International Contract Administration Techniques for Quality Enhancement-CATQEST 🖰

Pavements

European Asphalt Technology 🖰🖰
European Concrete Technology 🖰🖰
South African Pavement Technology
Highway Information Management
Highway/Commercial Vehicle Interaction

Bridges

European Bridge Structures
Asian Bridge Structures
Bridge Maintenance Coatings
European Practices for Bridge Scour and Stream Instability Countermeasures
Advanced Composites in Bridges in Europe and Japan 🖰

Planning and Environment

European Intermodal Programs: Planning, Policy and Technology 🖰
National Travel Surveys 🖰

Safety

Pedestrian and Bicycle Safety in England, Germany and the Netherlands 🖰
Speed Management and Enforcement Technology: Europe & Australia 🖰
Safety Management Practices in Japan, Australia, and New Zealand 🖰
Road Safety Audits – Final Report 🖰
Road Safety Audits – Case Studies 🖰
Innovative Traffic Control Technology & Practice in Europe 🖰
Commercial Vehicle Safety Technology & Practice in Europe 🖰

Operations

Advanced Transportation Technology 🖰
European Traffic Monitoring
Traffic Management and Traveler Information Systems
European Winter Service Technology
Snowbreak Forest Book – Highway Snowstorm Countermeasure Manual (*Translated from Japanese*)

Policy & Information

Emerging Models for Delivering Transportation Programs and Services
Acquiring Highway Transportation Information from Abroad – Handbook 🖰
Acquiring Highway Transportation Information from Abroad – Final Report 🖰
International Guide to Highway Transportation Information 🖰🖰

🖰 **Also available on the internet**

🖰🖰 **Only on the internet** at www.international.fhwa.dot.gov

Office of International Programs
Federal Highway Administration
400 Seventh Street, SW
Washington, DC 20590
tel: 202-366-9636
fax: 202-366-9626
http://international.fhwa.dot.gov/
email: international@fhwa.dot.gov

Publication NO. FHWA-PL-2000-010
HPIP/5-00(7M)EW

FHWA INTERNATIONAL TECHNOLOGY EXCHANGE REPORTS

Infrastructure

Geotechnical Engineering Practices in Canada and Europe 🖱
Geotechnology—Soil Nailing 🖱
International Contract Administration Techniques for Quality Enhancement-CATQEST 🖱

Pavements

European Asphalt Technology 🖱🖱
European Concrete Technology 🖱🖱
South African Pavement Technology
Highway Information Management
Highway/Commercial Vehicle Interaction

Bridges

European Bridge Structures
Asian Bridge Structures
Bridge Maintenance Coatings
European Practices for Bridge Scour and Stream Instability Countermeasures
Advanced Composites in Bridges in Europe and Japan 🖱

Planning and Environment

European Intermodal Programs: Planning, Policy and Technology 🖱
National Travel Surveys 🖱

Safety

Pedestrian and Bicycle Safety in England, Germany and the Netherlands 🖱
Speed Management and Enforcement Technology: Europe & Australia 🖱
Safety Management Practices in Japan, Australia, and New Zealand 🖱
Road Safety Audits – Final Report 🖱
Road Safety Audits – Case Studies 🖱
Innovative Traffic Control Technology & Practice in Europe 🖱
Commercial Vehicle Safety Technology & Practice in Europe 🖱

Operations

Advanced Transportation Technology 🖱
European Traffic Monitoring
Traffic Management and Traveler Information Systems
European Winter Service Technology
Snowbreak Forest Book – Highway Snowstorm Countermeasure Manual (*Translated from Japanese*)

Policy & Information

Emerging Models for Delivering Transportation Programs and Services
Acquiring Highway Transportation Information from Abroad – Handbook 🖱
Acquiring Highway Transportation Information from Abroad – Final Report 🖱
International Guide to Highway Transportation Information 🖱🖱

🖱 **Also available on the internet**

🖱🖱 **Only on the internet** at www.international.fhwa.dot.gov

Office of International Programs
Federal Highway Administration
400 Seventh Street, SW
Washington, DC 20590
tel: 202-366-9636
fax: 202-366-9626
http://international.fhwa.dot.gov/
email: international@fhwa.dot.gov

Publication NO. FHWA-PL-2000-010
HPIP/5-00(7M)EW

www.ingramcontent.com/pod-product-compliance
Lightning Source LLC
Chambersburg PA
CBHW081854170526
45167CB00007B/3009